Of Rocks, Mountains and Jasper

A Visitor's Guide to the Geology of Jasper National Park

EON	ERA	PERIOD		AGE IN MY.	JASPER MAIN RANGES FORMATIONS	JASPER FRONT RANGES FORMATIONS
PHANEROZOIC	MESOZOIC	TERTIARY		66		
		CRETACEOUS		135		LUSCAR Gp. ss, sh, coal
						CADOMIN Fm. cngl.
						NIKANASSIN Fm. ss,sltst,sh
		JURASSIC		208		FERNIE Gp. ss,sltst,sh,ch
		TRIASSIC		245		WHITEHORSE Fm. carb.
						SULPHUR MOUNTAIN Fm. sltst,sh
	PALEOZOIC	PERMIAN		286		ISHBEL Gp. ch,cngl,ss
		CARBONIFEROUS		360		RUNDLE Gp. carb.
						BANFF Fm. sh,sltst,carb.
						EXSHAW Fm. sh,sltst
		DEVONIAN	UPPER	374		PALLISER Fm. carb.
						SASSENACH Fm. ss
						SIMLA Fm. carb.
						MT. HAWK Fm. sh, carb. SOUTHESK Fm. carb.
						PERDRIX Fm. sh CAIRN Fm. carb.
						FLUME Fm. carb.
			MIDDLE	387		
			LOWER	408		
		SILURIAN		438		
		ORDOVICIAN		505		OUTRAM Fm. sltst,sh,carb.
		CAMBRIAN			SURVEY PEAK Fm. sh,/carb.	SURVEY PEAK Fm. sh,carb.
					LYNX Fm. carb.	LYNX Fm. carb.
					ARCTOMYS Fm. red & green sh	ARCTOMYS Fm. red & green sh
					PIKA Fm. carb.	PIKA Fm. ls
					TITKANA Fm. carb.	ELDON Fm. carb.
					CHETANG–TATEI Fm. carb.	SNAKE INDIAN Fm. sh,carb.
				570	GOG Gp. qtzite,carb.	
PRE-CAMBRIAN	PRO-TEROZOIC	UPPER		900	MIETTE Gp. slate,grit	OLDER STRATA NOT EXPOSED

Geological column and formations of sedimentary strata exposed in Jasper National Park and Mount Robson Provincial Park. The age, in millions of years ago, is shown for the boundaries between geological periods.

carb = carbonate
  (limestone and/or dolomite)
ch = chert
cngl = conglomerate
Fm = formation

Gp = group (of formations)
qtzite = quartzite
sh = shale
sltst = siltstone
ss = sandstone

Of Rocks, Mountains and Jasper

A Visitor's Guide to the Geology of Jasper National Park

Chris Yorath and Ben Gadd

Dundurn Press
Toronto • Oxford

Copyright © Minister of Supply and Services Canada 1995

All rights reserved. No part of this publication may be reproduced, stored in a retrieval system, or transmitted in any form or by any means, electronic, mechanical, photocopying, recording, or otherwise (except brief passages for purposes of review), without the prior permission of Dundurn Press Limited. Permission to photocopy should be requested from the Canadian Reprography Collective.

Designed by Andy Tong
Edited by Judith Turnbull
Printed and bound in Canada by Best Book Manufacturing

The publisher wishes to acknowledge the generous assistance of the **Geological Survey of Canada** and the ongoing support of the **Canada Council**, the **Book Publishing Industry Development Program** of the **Department of Canadian Heritage**, the Ontario Arts Council, the Ontario Publishing Centre of the **Ministry of Culture, Tourism and Recreation**, and the Ontario Heritage Foundation.

Care has been taken to trace the ownership of copyright material used in the text (including the illustrations). The author and publisher welcome any information enabling them to rectify any reference or credit in subsequent editions.

J. Kirk Howard, Publisher

Canadian Cataloguing in Publication Data

Yorath, C.J., 1936–
  Of rocks, mountains and Jasper : a visitor's guide to the geology of Jasper National Park

Co-published by the Geological Survey of Canada.
Includes index.
ISBN 1-55002-231-8

1. Geology – Alberta – Jasper National Park – Guidebooks. 2. Jasper National Park (Alta.) – Guidebooks. I. Gadd, Ben, 1946– . II. Geological Survey of Canada. III. Title.

QE186.Y67 1995   557.123'32   C95-930779-6

Dundurn Press Limited	Dundurn Distribution	Dundurn Press Limited
2181 Queen Street East	73 Lime Walk	1823 Maryland Avenue
Suite 301	Headington, Oxford	P.O. Box 1000
Toronto, Canada	England	Niagara Falls, N.Y.
M4E 1E5	OX3 7AD	U.S.A. 14302-1000

Contents

COLOUR PLATES	83–98
PREFACE	ix
How the book is organized	ix
Acknowledgments	xii

PART ONE
UNDERSTANDING THE GEOLOGY OF THE CANADIAN ROCKIES   1

HOW OLD IS THAT MOUNTAIN?	6
A SUPERCONTINENT IS TORN APART	6
THE BEGINNINGS OF JASPER	8
A SHALLOW INLAND SEA	10
RICHES BEQUEATHED BY CORALS AND STROMATOPOROIDS	12
ANOTHER SUPERCONTINENT COMES TOGETHER, THEN RIFTS APART	13
A VOLCANO HERE AND THERE	17
CRASH. BANG. CRUNCH.	19
WHAT OF THE FOOTHILLS?	22
TEN KILOMETRES OF ROCK GONE DOWN THE RIVER	26
ICE, THE GREAT SCULPTOR	30
THE MOUNTAIN LANDSCAPE WE SEE TODAY	34
MORE ABOUT THE GEOLOGICAL MAP	39

Part Two
A Guide to the Geology of Jasper National Park 43

The Yellowhead Highway (Highway 16) Jasper to Tête Jaune Cache 44

- ① Alberta/British Columbia border: the Continental Divide — 46
- ② Yellowhead Lake — 47
- ③ Mt. Fitzwilliam — 47
- ④ Moose Lake — 49
- ⑤ Mt. Robson — 49
- ⑥ Selwyn Range — 51

The Yellowhead Highway (Highway 16) Jasper to the Mountain Front 53

- ⑦ Boundary between the main ranges and front ranges — 57
- ⑧ The Palisade — 58
- ⑨ Folds and faults in the front ranges — 59
- ⑩ River Rock and Cold Sulphur Spring — 60
- ⑪ Jasper Lake — 64
- ⑫ Roche Miette — 67
- ⑬ Roche à Perdrix — 67
- ⑭ The mountain front — 67

The Fiddle Road Punchbowl Falls to Miette Hotsprings and the Sulphur Skyline Trail 70

- ⑮ Punchbowl Falls — 70
- ⑯ Ashlar Ridge — 72
- ⑰ Fernie Formation at Morris Creek — 74
- ⑱ Miette Hotsprings — 76
- Sulphur Skyline Trail — 79

The Maligne Valley 81

- ⑲ Maligne Canyon — 81
- ⑳ Maligne River and Medicine Lake — 102
- ㉑ Maligne Lake — 104
- The Skyline Trail — 108

Note: Circled numbers correspond to numbers on geological map (in pocket on inside back cover) indicating points of special interest.

The Icefields Parkway (Highway 93)
Jasper to Sunwapta Pass — 115

- ㉒ Athabasca Falls — 115
- ㉓ Goats and Glaciers Viewpoint — 115
- ㉔ Sunwapta Falls — 119
- ㉕ Endless Chain Ridge, Bubbling Springs and the Jonas Slide — 119
- ㉖ Beauty Creek Flats — 121
- ㉗ Tangle Ridge — 123
  - Viewpoints on the Tangle Hill — 124
    - ㉘ Stutfield Viewpoint — 124
    - ㉙ Tangle Falls Viewpoint — 124
    - ㉚ Mount Kitchener Viewpoint — 124
- ㉛ Athabasca Glacier and the Columbia Icefield — 126
- ㉜ Sunwapta Pass and Nigel Peak — 131

Other Localities of Geological Interest — 133

- ㉝ Jasper townsite area — 133
- ㉞ Marmot Mountain — 137
- ㉟ Pyramid Mountain — 137
- ㊱ Old Fort Point — 138
- ㊲ The Whistlers — 142
- ㊳ Mt. Edith Cavell — 144
  - The Tonquin Valley via Astoria River and Maccarib Pass — 149

Glossary — 153
Sources and Additional Reading — 164
Index — 166

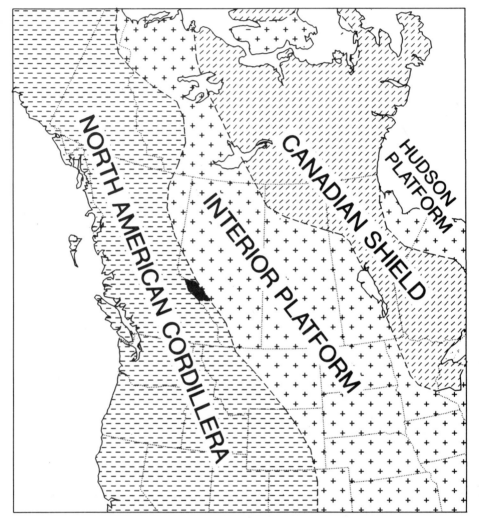

Principal geological/physiographic subdivisions of northwestern North America and the approximate location of Jasper National Park.

Preface

This book is about rocks, mountains and Jasper. It's about thrust faults and colliding continents. About glaciers and rivers, mountain goats and landslides. It is a story of time and the building of western Canada. And here, in Jasper National Park, part of that story is told by the rocks of these magnificent mountains.

HOW THE BOOK IS ORGANIZED

The book is organized as two parts. The first describes how geologists think these mountains came to be. We describe what they are made of, when they were made and how. Here we talk about the origin of the **Canadian Cordillera,** the system of mountain ranges and plateaus extending from the International Boundary to the Arctic Ocean, and from the foothills of the Rockies to the edge of the continent off the west coast of Vancouver Island and the Queen Charlottes. The second part is more specific to the Jasper area. It forms a guide to the geological phenomena in the park. Here we talk about the rocks and mountains of Jasper as they are seen from vantage points along the highways and from some of the more popular hiking trails.

In a pocket at the back is a geological map that illustrates the general **physiography,** or landforms, of Jasper National Park, as well as the age, kinds and distribution of the rocks that form its mountains. The map shows physiographic features in shaded relief. The result is a three-dimensional view of the park as seen in the morning, with the sun shining from the east. The map shows the geological architecture of the park as if it were stripped of its forests, surface soil, glacial deposits, sands and gravels. Each colour represents sedimentary rocks that accumulated during specific intervals of geologic time. For example, purple represents those areas at the surface with outcrops of strata deposited during the latter part of the Devonian Period of the Paleozoic Era, between 375 and 360 million years ago; dark brown shows those sediments that were deposited during the latter part of Precambrian time, between about 750 and 590 million years ago. There are many other colours and symbols on the map, and they are explained in its legend.

The graph at the front of the book, facing the title page, shows the geological time scale, including the ages in millions of years ago of the boundaries between geological periods and eras. In a parallel column we illustrate the succession of major sedimentary rock layers, or **formations,** that occur in Jasper National Park, as well as in adjoining Mount Robson Provincial Park in British Columbia. By referring to the caption you can get an idea of the main rock types in these formations. The position of each formation in the column is in accordance with its age. The patterned gaps between formations indicate intervals of time for which rocks are not found in the two parks. Throughout the book we frequently refer to these geological formation names and time terms.

We have attempted to keep geological jargon to a minimum. Each technical term is defined the first time it is used, and a glossary at the back will further help the reader.

PREFACE / XI

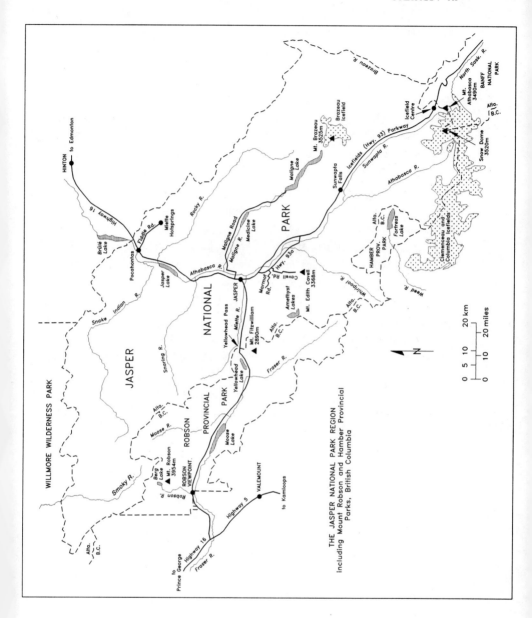

Acknowledgments

All of the information contained in this book was obtained from reports, published and unpublished, and from verbal communications with other geologists. Of particular importance are the scientific publications of Eric Mountjoy of McGill University, Ray Price of Queen's University, and Henry Charlesworth and his students at the University of Alberta. In association with geologists of the Geological Survey of Canada (GSC), Eric and Ray undertook the detailed mapping of Banff, Yoho, Kootenay and Jasper national parks under the auspices of the GSC's "Operation Bow-Athabasca" in 1965 and 1966. This was a helicopter-supported geological mapping project during which vast quantities of data were obtained on rock types and their ages, distribution and structure. From these and subsequent studies over the past three decades our current understanding of the geological history and development of the Rockies emerged. We also acknowledge the kindness of Shell Canada Resources Ltd., who provided us with their geological maps of the Maligne Lake region. The geological map and drawings were done by Brian Sawyer and Richard Franklin of the GSC. Don Cook and Carmel Lowe of the GSC and Dick Campbell (GSC-retired) provided valuable critical comment that improved the final manuscript. Linda Yorath, Cia Gadd, Arthur Walters, Marlaine Brown, Lawrie Law and Doug McLean read the manuscript for clarity. To all of these people we express our sincere thanks.

Geologists like to stand with their backs to precipices. Eric Mountjoy (above), professor of geology at McGill University, stands on Paleozoic rocks high above the south end of Maligne Lake. Henry Charlesworth, professor of geology at the University of Alberta, is standing on Cambrian strata above the valley of the Miette River. It is to these men, their students and their colleagues that we owe our appreciation of the geology of Jasper National Park.

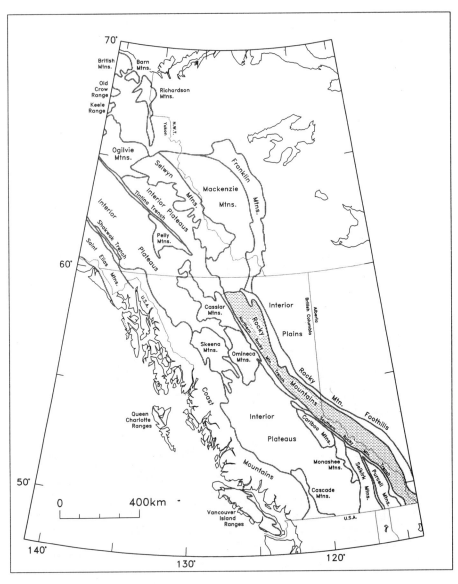

Physiographic map of the Canadian Cordillera. The several mountain systems occur as northwest/southwest-aligned groups of ranges, each having markedly different character, depending upon their origin and rock composition. The Rockies (stippled), Mackenzies, Ogilvies and Richardson Mountains are composed of sedimentary rocks, whereas most of the other systems have been formed from sedimentary, igneous and metamorphic rocks. A broad system of plateaus separates the mountain systems of the eastern and western Cordillera.

PART ONE

UNDERSTANDING THE GEOLOGY OF THE CANADIAN ROCKIES

At the very beginning it should be understood that the Canadian Rockies form only a part of the mountain systems of western Canada. Whereas some people believe that the Rockies extend from the foothills of Alberta and British Columbia to the west coast, such is not the case. The **Canadian Cordillera*** consists of many mountain systems, including the Rockies, Purcell Mountains, Selkirks, Cariboos, Skeena, Coast, St. Elias and many others. As shown on the map (facing page), the Rocky Mountains include those ranges lying between the foothills and the Rocky Mountain Trench. Forming the easternmost mountain system of the southern Canadian Cordillera, the Canadian Rockies extend from the international boundary to the Liard River, close to the northern boundary of British Columbia, beyond which lies the broad arc of the Mackenzie Mountains in the Northwest Territories and Yukon.

The exposed rocks in the Rocky Mountains of Jasper National Park are almost everywhere sedimentary rocks. Despite the assertions of those who speak of "the granite wall of the Canadian Rockies," with rare exceptions there are virtually no igneous rocks, no granites or lavas, exposed in the Canadian Rockies. We say "exposed" because granitic rock forming the westward-sloping surface of the Precambrian **Canadian Shield** is

---

*Where we introduce a technical term, we use **boldface** type. In most cases, these terms are introduced in an explanatory context and are defined in the glossary, pages 153–63.

estimated to occur about 9 km below the surface of the park. So, while granite does exist in the area, it is deeply buried beneath a thick cover of layered sedimentary rocks.

Another word for sedimentary layers is **strata**. Looking up at the cliffs throughout the park, you will see that most are stratified or banded in different colours or shades. Some layers are thick; others are thin. These differences between one bed, or **stratum**, and another are caused by variations in one or more of a wide variety of factors, including the physical, chemical and biological environment that existed at the time the sediments accumulated.

The sedimentary strata of the park are broadly of two main types: **clastics** and **carbonates**. **Clastic** rocks are composed of particles, such as grains, or **clasts**, of sand that have been eroded from other rocks and transported to the place where they have been deposited by mechanical agents such as water, wind, ice or gravity (landslides, etc.). Examples of clastic rocks include **sandstone, shale and conglomerate.** In Jasper, the minerals that most commonly comprise the grains of clastic rocks are **quartz** and **feldspars.** Quartz is a hard, insoluble (at surface temperatures and pressures) compound of silica and oxygen, capable of withstanding many cycles of erosion and deposition, whereas minerals such as feldspars are easily broken down during these processes and converted to **clays,** which commonly fill the spaces between the larger clasts.

Sandstone is a comparatively soft clastic rock made almost entirely of clasts of quartz. If the individual quartz grains are held together by microscopic crystals of quartz, the rock is much harder and is called **quartzite.** Many of the high peaks of the western part of the park, such as Mt. Edith Cavell, are formed from quartzite. **Conglomerate** is essentially

Finely laminated layers of quartzite of the Cambrian Gog Group consist of grains of quartz and lesser amounts of feldspar, tightly cemented together by silica.

Photo by Ben Gadd

gravel turned into rock, composed of clasts of a variety of rock types greater than 2 mm in diameter. Conglomerate looks rather like stony concrete. **Shale** is rock-hard mud, which normally consists of fine clasts of silt and clay.

Carbonates are sedimentary rocks of calcium, magnesium or iron carbonates – minerals that include a combination of carbon (thus the term "carbonate") and oxygen. Carbonates usually are deposited in seawater. The most ubiquitous carbonates in Jasper are **limestone** and **dolomite**, the former composed of calcium carbonate ($CaCO_3$; the mineral is called **calcite**) and the latter of calcium and magnesium carbonate ($Ca,MgCO_3$; the mineral, like the rock, is called **dolomite**).

All carbonates in the Rockies were formed through organic means, that is, through the ability of marine organisms (e.g., algae, corals and clams) to extract calcium carbonate from seawater to build their colonies, shells and skeletons. Surprisingly, most limestone originates from tiny crystals of lime formed within the cells of cyanobacteria, formerly called "blue-green algae." The limestones and dolomites of the Rockies are mainly composed of recrystallized lime mud (the result of the rupture of cyanobacteria cells), with the addition of fragmented shells of marine animals. In some cases whole shells of fossil organisms can be seen in the rocks; more often, however, the shells have been broken up by scavenging organisms and the abrasive action of marine bottom currents to such an extent that only tiny fragments remain. A good example of fossiliferous limestone can be seen at Jasper Park Lodge, where the outside stairways and hotel entrance are paved with the well-known Tyndall limestone of Manitoba. This limestone contains easily recognized large fossils of corals and other sea life imbedded in calcium carbonate.

Although most of the carbonate rocks of the Rockies originally formed as limestone, some have been converted to dolomite. The process

Dark grey- to black-weathering layers of sandy limestone are interstratified with buff-coloured beds of silty dolomite in this sample from the Cambrian Lyell Formation.

Photo by Ben Gadd

Angular fragments of carbonate, shale and chert are enclosed within coarse-grained sandstone to form the Cretaceous Cadomin conglomerate.

Photo by Ben Gadd

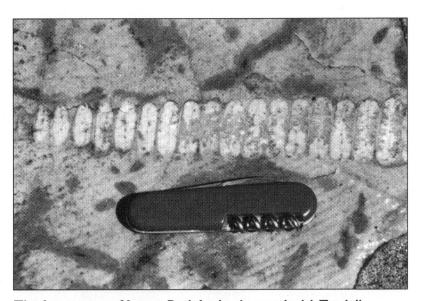

The front entry of Jasper Park Lodge is paved with Tyndall limestone slabs quarried in Manitoba. This popular building stone contains many different kinds of fossils, such as cephalopods (shown here) and corals.

Photo by Ben Gadd

of **dolomitization** of limestone involves the partial substitution of calcium by the element magnesium in the atomic structure of calcite. During the process, whole fossil remains of organisms commonly are destroyed or replaced by other minerals. Furthermore, dolomitization leaves voids between the mineral grains. Deep beneath the Alberta plains, fluids such as oil and gas are trapped within these porous, sponge-like rocks, thus forming many of Canada's oil and gas fields.

Many of the rocks in the western part of the Rockies are **metamorphic** rocks. Originally consisting of sandstones and shales, these were later subjected to intense heat and pressure through deep burial and/or forces involved in mountain-building. The effect was to change the mineral make-up of the rocks. Some of the clay minerals in shale, for example, were changed to flakes of mica and chlorite; the rock is now harder and tends to split across the stratification rather than along it. These rocks we call **slate**. Not far west of Jasper National Park we find silvery **schist**, in which the applied heat and pressure were greater than in the creation of slate, changing the original minerals to mica, **garnets** and other metamorphic minerals indicative of a more intense degree of metamorphism. **Gneiss**, a banded rock formed under even greater heat and pressure, is the highest grade of metamorphic rock known in the Canadian Rockies. Gneiss is found on the east side of the Rocky Mountain Trench southwest of Jasper National Park.

Thin slabs of slate are characteristic of many outcrops of the Precambrian Miette Group along the Yellowhead Highway west of Jasper.

Photo by Ben Gadd

How old is that mountain?

To understand the geology of the Rockies it is important to realize that *rocks come first and mountains second*. By this we mean that the rocks you see in the cliffs of Mt. Edith Cavell were formed some 570 million years ago but elevated high above sea level beginning about 100 million years ago. A mountain can be compared to a house: while the wood used to build the house may come from hundred-year-old trees, the house itself may have been constructed only last month. As rocks go, those of the Rockies are fairly old. As mountain ranges go, the Canadian Rockies are middle-aged: mountain-building here occurred well after the building of old ranges such as the Appalachians but well before that of the Himalayas and St. Elias Mountains, which are still rising today.

The second thing to realize is that *as the mountains rose, erosion modified their shapes and wore them down*. We have mountains here today only because the rate of mountain-building exceeded the rate of erosion. Over the 120 million years that have elapsed since the Canadian Rockies began to rise above sea level, a great deal of rock has been removed by erosion. It is important to understand that the actual shapes of the mountains and valleys in the park are the result of the enormous sculpting power of water and glacial ice. The mountains have not simply been pushed up and left in the configuration that their folds and faults gave them. Rivers have been modifying their shapes ever since the area rose above sea level. Glaciers have covered the entire Canadian Cordillera off and on for the past 2 million years, carving deeply into the ranges, giving them a very different look than they had before the ice ages. The Columbia Icefield and other glaciers throughout the park and elsewhere in the Cordillera are, in a sense, the last remnants of that ice, still scraping away at the peaks. This means that, geologically speaking, the landscape you see is young.

These, then, are the key ideas to keep in your mind as you read this book: the age of the rock (old), the age of mountain-building (moderate), and the age of the landscape (young).

A supercontinent is torn apart

The origin of the Canadian Rockies is intimately linked to that of other mountain systems of the Canadian Cordillera, and linked as well to the geological architecture and history of the Interior Plains. To let you gain a feeling for these linkages, we will take you back to a time when there were no eyes to see, no ears to hear.

Imagine that you are standing at the site of Jasper as it was about 750 million years ago. A pallid sky surrounds a yellow moon that appears much larger than it does today — larger because it is closer. The days are shorter because the earth spins faster. The air is dry and cool. The Precambrian ground under you feet is swept by wind-blown eddies of sand carried from high mountain ranges to the east. There are no trees, no grass, no animals — no life on the land at all. Farther west are other mountain ranges, which today are believed to have formed part of continental Australia. Yes, Australia.

The supercontinent of Rodinia as it might have looked prior to 750 million years ago.

From an illustration by P.F. Hoffman

Some geologists believe that by about 1750 million years ago, during the Precambrian Eon, the world's continents had become amalgamated into one giant supercontinent called **Rodinia.** That part of the supercontinent that was to become Australia lay next to western North America. Between the two was a small, trapped ocean basin, much like the Black Sea, which for about a billion years had been receiving sediments eroded from the surrounding mountainous continents. Those sediments, now hardened into rock, can be seen today in the mountains of Waterton Lakes National Park at the southern end of the Canadian Rockies. There is no evidence of these very old sediments in the Jasper area; they either were not deposited here or, if they were, have since been eroded away.

About a billion years after its formation, the supercontinent of Rodinia began to break up. (This was the point at which we put you, standing in an alien and lifeless Jasper.) Over the next several million years, Australia rifted away from the rest of Rodinia in the same way that North America split apart from Africa some 550 million years later, when the modern Atlantic Ocean was created. The **rift** (the line of splitting) through Rodinia, which created a new edge for western North America, developed more or less along a line joining the sites of the modern communities of Penticton, Kelowna, Prince George and points north and south. As Australia moved away, a broad ocean basin formed between the separating continents. The stage was set for accumulation of the oldest rocks in Jasper National Park.

THE BEGINNINGS OF JASPER

Immediately following separation, a thick pile of sediments began to accumulate across the torn continental edge. As more and more sediment was eroded from the granitic continental interior and carried by rivers to the ocean, a continental shelf formed in much the same way as the Grand Banks of Newfoundland would many millions of years later. In the Jasper region these sediments can be seen in outcrops and roadcuts beside the Yellowhead Highway west of the town, where the road follows the Miette River. These rocks have been named the Miette group of **formations,** or, more simply, the Miette **Group.** The thickness of the Miette Group is astonishing: about 10 km! Much of it consists of alternating layers of **gritstone** (sandstone, with the addition of grains of feldspar and mica) and slate.

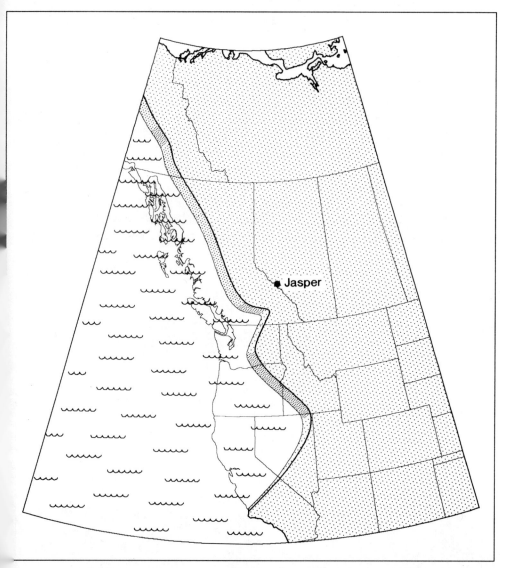

The edge of western North America about 750 million years ago, after Australia and East Antarctica had rifted away.

A SHALLOW INLAND SEA

Let us return to our Jasper vantage point. One hundred and eighty million years have passed since North America separated from Australia. It is now the Cambrian Period of the Paleozoic Era (roughly 570 million year ago). The mountain ranges of the Canadian Shield have eroded away; rivers and winds have carried the eroded fragments westward to the sea, which covers a wide continental shelf. The Precambrian granite surface, once underlying the seabed, now lies buried beneath the Miette Group and an additional 2–3 km of Cambrian sandstone and siltstone called the **Gog Group.**

Because of the accumulated sediments, you are now several kilometres above the actual old site of 750 million years ago. Hopefully you have developed gills, for you now find yourself standing on the floor of a shallow

sea covering the continental shelf. The water is less than 150 m deep.*
Whereas throughout the hundreds of millions of preceding Precambrian years there was little life except for algae, bacteria and a few primitive worm-like and jellyfish-like creatures, now the oceans have begun to teem with exotic-looking sea life. You can see flat, oblong, segmented creatures with bulbous eyes skittering about the seabed on their many legs. These are **trilobites,** with names such as *Oleanullus, Fremontia* and *Nevadella.* They leave their tracks and trails in the Gog limy muds, and when they die, their bodies are entombed in the mud to become preserved as fossils.

As the millennia roll by, the sea advances ever farther inland, finally reaching well into the interior of the continent. In shallow areas where the waters are warm and clear, carbonate reefs develop. Organisms called **archaeocyathids** are the main reef-builders; their carbonate shells look rather like ice-cream cones. In deeper waters farther offshore, trilobites still slither across the muddy seabed, their insect-like shells becoming buried by the constant rain of fine limy mud.

Today, Lower Cambrian rocks are spectacularly well exposed in mountains throughout Jasper National Park. Mounts Kerkeslin, Edith Cavell and Tekarra, the Watchtower, Pyramid Mountain, as well as the massive wall of the Ramparts above the Amethyst Lakes in the Tonquin Valley, display the Gog Group sandstones, now hardened to quartzite. To the west, in Mount Robson Provincial Park, Mt. Robson (the highest mountain in the Canadian Rockies at 3954 m) displays a magnificently photogenic succession of Middle and Upper Cambrian strata capped by a white crown of snow and ice, where turbulent clouds can often be seen roiling about (see photo on page 85).

Throughout the Ordovician and early part of the Silurian periods, between 500 and 420 million years ago, living things continued to increase both in numbers and diversity. **Brachiopods,** corals and **bryozoans** thrived upon the seabeds, while organisms called **graptolites,** now long extinct, floated about on the surface. Today the remains of graptolites are found in black shales, where they resemble tiny hacksaw blades.

During these 80 million years in western Canada, the seas advanced and retreated many times in concert with the slow, irregular upward and downward movements of the earth's crust. As well, the sea level rose and fell globally. At times the surf broke as far east as Hudson Bay, while at other times and places the seabed became exposed to the rain and wind,

---

*You might wonder how 10 km of sedimentary strata could have been deposited in just 150 m of water. As the sediments accumulated, the earth's crust sank under the additional weight of the sediment load. Also, following the rifting phase, the edge of the continent cooled and thus sank as it became more dense.

which eroded the rock of the emergent land surface. These sediments were carried far to the west, to be redeposited in deep waters along the rim of the continent.

During the latter part of the Silurian Period and the early part of the Devonian Period, marine waters receded over much of western Canada, exposing the seabed to erosion. The record of 85 million years of geological history was lost in the Jasper area as successively deeper layers were eroded away. But the sea readvanced again during Late Devonian time, covering the old erosion surface with a new blanket of carbonate strata.

RICHES BEQUEATHED BY CORALS AND STROMATOPOROIDS

The latter part of the Devonian Period saw the formation of many of our important oil and gas reservoirs. The GSC's Helmut Geldsetzer has explained how, on the floor of a shallow, clear and warm sea with broad carbonate banks, local circulation and bottom topographic conditions allowed for the rapid growth of linear chains of reefs constructed by **stromatoporoids**, algae and corals. Between the reefs, fine muds accumulated, containing abundant organic material. Following burial and hardening, the increase in heat cooked the organic material into petroleum, which then migrated into and became entrapped within the porous, sponge-like reefs. Oil fields such as Leduc, Redwater and Swan Hills now lie deeply buried beneath the Alberta plains.

In the more remote regions of Jasper National Park, these same kinds of reefs are exposed in the Ancient Wall along Blue Creek in the Snake Indian River region in

Dissolved away during the process of conversion from calcite to dolomite, the centres of stromatoporoids provide a porous structure that holds oil and gas within Devonian reefs deep below the grain fields of Alberta.

Photo by Ben Gadd

the northern part of the park, and in the Miette Range southeast of Miette Hotsprings, where they have been studied by Eric Mountjoy of McGill University (see illustration on page 14). Eric has worked out the several stages of growth of the Miette reef complex, a growth that we estimate to have occurred over an interval of some 3 million years, beginning about 375 million years ago.

ANOTHER SUPERCONTINENT COMES TOGETHER, THEN RIFTS APART

During the interval when Jasper's Devonian and Carboniferous strata were accumulating on the shallow seafloor, the world's continents were again coalescing into another giant supercontinent. This one we call **Pangea**. By about 300 million years ago, in late-Carboniferous time, Pangea had assembled. The supercontinent lasted for about 100 million years, until the beginning of the Jurassic Period of the Mesozoic Era, after which it, too, began to break up. First it separated into two large chunks: a northern mass called **Laurasia**, which included North America and most of Europe and Asia, and a southern mass called **Gondwanaland**, including South America, Africa, Antarctica, Australia and India. The globe-encircling ocean surrounding Pangea is called **Panthalassa**; as a result of the initial breakup, the Panthalassan ocean extended between the separating masses to form the North Atlantic Ocean.

The process by which continents separate is called **seafloor spreading**, a component of the theory of **plate tectonics**. Molten lava, originating from the **mantle**, the soft, hot layer beneath the earth's crust, is injected into long fissure systems known as **spreading ridges**. Upon encountering cold seawater, the lava rapidly cools, sticking to the sides of the fissure system. With each succeeding injection, the walls of the fissure system move apart to make way for the new material; by this means the rocks of the seafloor are created and are spread outward away from the ridges. When a spreading ridge forms beneath a continent, it can force the continent to split, or rift, apart, thus allowing the two pieces to move away from each other. Oceanic crust fills the gap, as it did along the northern mid-Atlantic Ridge when North America (Laurasia) rifted away from Eurasia. In the same manner, Australia was torn away from North America some 550 million years earlier.

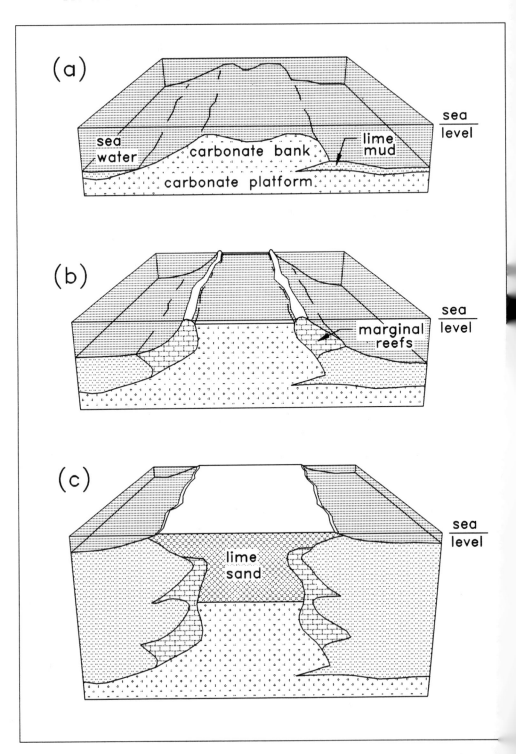

◀ This model showing three stages of growth of the Devonian Miette reef complex is based upon studies by Eric Mountjoy of McGill University. In *(a)*, a broad carbonate platform (Cairn Formation) was constructed by stromatoporoids in a clear, warm, shallow sea that spread over the western Canada shelf about 375 million years ago. As the sea deepened, stromatoporoid growth accelerated in areas where the underlying platform was shallowest and thickest, resulting in the development of a carbonate bank. The action of waves caused erosion of the outer parts of the bank, resulting in the accumulation of shaly carbonate (Maligne Formation) in surrounding deeper water. In *(b)*, continued deepening caused further restriction of stromatoporoid bank growth and the development of coral and stromatoporoid reefs around the bank margins. These marginal reefs, which emerged above sea level in some locations, were highly porous, allowing magnesium-rich waters to convert the limestone to dolomite (see photo on page 12). In the flanking basin, black muds (Perdrix Formation) accumulated in a stagnant environment surrounding the reef complex. In *(c)*, further growth of marginal reefs, largely consisting of corals, continued after the sea level had stopped rising. Waves attacked the bordering reefs, resulting in the accumulation of lime sands (Southesk Formation) in lagoons behind the reefs. In flanking deeper water, lime muds (Mount Hawk Formation) were deposited from currants carrying debris eroded from distant sources. If we assume rates of reef growth equivalent to those in modern oceans, the Miette reef complex developed during an interval of about 3 million years.

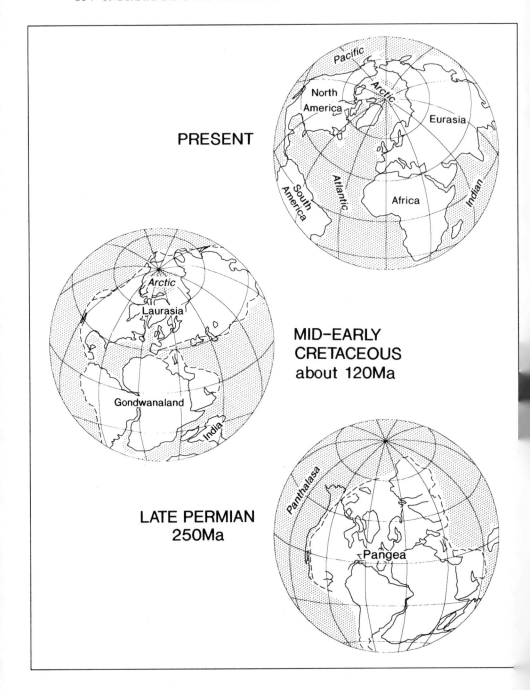

◀ Three stages of the opening of the Atlantic Ocean. During the Permian Period, about 250 million years ago, the globe's continents were gathered together into one giant supercontinent called Pangea. Beginning about 180 million years ago, Pangea disassembled into two huge continental masses separated by a widening Atlantic Ocean. To the north was Laurasia, consisting of North America and Eurasia, and to the south was Gondwanaland, made up of South America, Africa, Antarctica, Australia and India. The present configuration of the continents is shown at the top; India was added to Eurasia about 40 million years ago.

Based on drawings by E. Irving

A VOLCANO HERE AND THERE

The separation of Laurasia from Gondwanaland, resulting in the opening of the North Atlantic, occurred about 180 million years ago, just as the dinosaurs were approaching their heyday in western Alberta. Since then, the motions of the Western Hemisphere's continents have been ever westward. Ultimately, perhaps about 150 million years from now, western North America will crash into the eastern edge of Asia to form another supercontinent, which Paul Hoffman, a geologist at Harvard University, suggests might be called *Amazia!*

It was during this period of westward drift that much of British Columbia was added to the continent. The mechanism by which this was accomplished – called **subduction** – is again related to seafloor spreading and plate tectonics. For each cubic metre of seafloor created at the globe's oceanic ridge systems, an equivalent volume must be consumed somewhere else; if this were not so, our planet would be getting larger, which seems not to be the case. The consumption of oceanic crust occurs along deep-sea trenches, most of which are found bordering the continents facing the Pacific Ocean. As North America and South America move westward, the oceanic crust of the Pacific, created along the Juan de Fuca Ridge and East Pacific Rise, is consumed, or subducted, beneath the continents. Volcanic islands and limestone reefs carried upon the oceanic crust are thus scraped off and added to the continent when their supporting seafloor is subducted.

A look at a map of the Pacific shows countless islands and coral **atolls**. Each of these is moving along on the piece of Pacific seafloor upon which it rests. For the most part these islands are volcanic, formed from molten material rising upward from the underlying hot mantle, into and through the oceanic crust. During the process of being built, a volcanic edifice such as this forms what is called a **seamount** below the sea surface, and then, when it reaches the sea surface, it becomes a volcanic island, commonly surrounded by coral atoll reefs. In many cases the volcanic piles

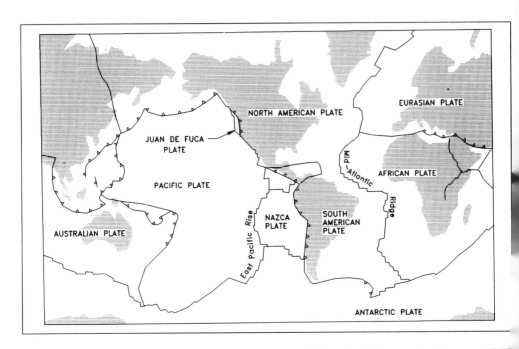

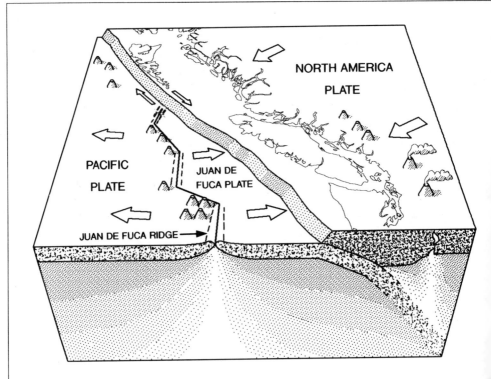

grow as chains of islands, the chains having an arcuate, or curved, shape. We call these chains **volcanic island arcs,** modern examples being the Aleutians, the Marianas, the Solomons and Indonesia. These are the kinds of things — island arcs and atolls, as well as pieces of the deep seafloor, assembled into fragments of crust called **terranes** — that make up much of British Columbia and the Yukon.

During the time when Devonian reefs were growing in warm, clear, shallow seas covering the western slope of the continent, and when Carboniferous, Permian and Triassic rocks were accumulating, terranes were being formed somewhere out in the vast Panthalassan ocean. By the late Paleozoic and early Mesozoic eras, several large terranes had assembled, probably near the equator. Through the processes of seafloor spreading and plate tectonics, these terranes were transported northward. Along the way they amalgamated with other terranes to form **superterranes,** which ultimately collided with the westward-moving continent. It was the force of these collisions that led to the formation of our Rocky Mountains.

CRASH. BANG. CRUNCH

There were two main collisional episodes. The first occurred when North America crashed into the **Intermontane Superterrane,** consisting of several smaller terranes made up of many groups of volcanic island arcs and coral atolls, as well as chunks of deep seafloor composed of **radiolarian oozes,** muds, lavas and even pieces of the mantle torn loose during the collisions. This event occurred during the middle part of the Jurassic Period, about 170 million years ago. Today, the Intermontane Superterrane

---

◀ The earth's outer shell, including the crust and upper part of the underlying mantle, is divided into several independently moving plates that are separated from one another by seafloor-spreading ridges, faults or deep-sea trenches (top). The seafloor-spreading ridges, such as the Mid-Atlantic Ridge, East Pacific Rise and Juan de Fuca Ridge, are zones along which new seafloor is created and added to the edges of plates. As new material forms, the plates spread away from the ridges toward the deep-sea trenches, where they are consumed, or subducted, beneath adjacent plates. The current plate tectonic situation off the west coast is shown in the lower figure. Molten magma from the mantle is injected into the Juan de Fuca Ridge, creating the Juan de Fuca Plate to the east and the Pacific Plate to the west. As the Juan de Fuca Plate spreads eastward, it is subducted beneath the westward-moving North American Plate. At a depth of between 150 and 200 km the oceanic crust melts, whereupon the molten material rises upward to appear at the surface of the overriding plate as a volcanic arc.

occupies most of central British Columbia and part of the southwestern Yukon.

The second event happened about 70 million years later, when the **Insular Superterrane** collided with the former superterrane (the Intermontane Superterrane), which by then had become welded to the continent. The Insular Superterrane consists of rocks found at scattered localities throughout the Coast Mountains; more importantly, it makes up most of the Paleozoic to Jurassic rocks of Vancouver Island, the Queen Charlotte Islands, southwestern Yukon and the Alexander Archipelago of southeastern Alaska.

As you can imagine, the effects of these collisions were traumatic indeed. The enormous heat and pressure caused the colliding edges to melt and deform. The western part of the thick apron of strata that had accumulated across the torn edge of the old continent was metamorphosed and intruded by molten magma derived through partial melting of the mantle below the crust. In some places the eastern edge of the colliding terrane was shoved up onto the western edge of the continent — a process called **obduction,** the opposite of subduction, whereby the edge

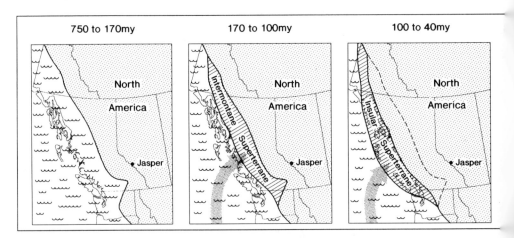

From the time of the breakup of Rodinia, about 750 million years ago, to the middle part of the Jurassic Period, western Canada consisted of a broad continental shelf, periodically covered by shallow tropical seas in which carbonate strata accumulated upon the shelf (left). About 170 million years ago the westward-moving continent collided with a large fragment of northward-moving crust and mantle called the Intermontane Superterrane, itself an amalgamation of several smaller terranes (middle). Approximately 70 million years later the continent collided with the Insular Superterrane (right), following which, beginning about 40 million years ago, several smaller terranes stuck onto Vancouver Island and western Alaska.

of the continent was depressed beneath the weight of the newcomer. To the east the apron became detached from its Precambrian granitic foundations and, like a thick stack of rugs being pushed across a hardwood floor, was shoved eastward, wrinkling into folds as it went. Not only did the stack accordion, it broke into **thrust sheets:** huge, thick slabs that overrode one another and overlapped like shingles on a roof. The breaks between thrust sheets are called **thrust faults.** (See illustration in glossary, page 162.) The force of the two collisions continued until about 60 million years ago, thrusting the sheets farther and farther eastward, some as much as 250 km, to where they are seen today in the Jasper Rockies.

Driving eastward along the Yellowhead Highway from Mt. Robson to Roche Miette, you can see the differing effects of these collisions on the rocks that make up the Rockies. High up on the face of Mt. Robson the carbonate strata are slightly bent or down-folded, forming a structure called a

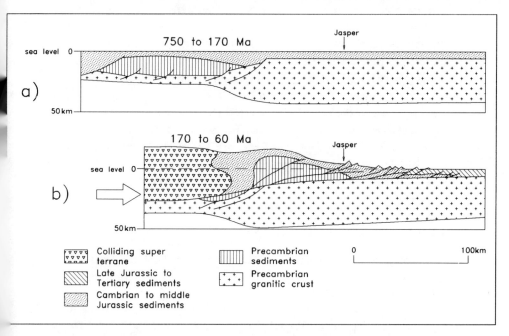

Following the breakup of the supercontinent Rodinia, western Canada consisted of a broad continental shelf that had developed across the torn and rifted edge of the continent. On the surface of the shelf a westward thickening wedge of strata accumulated throughout the Paleozoic and early part of the Mesozoic eras (a). The collision of North America with the superterranes had the effect of compressing the sedimentary wedge, breaking it loose from the granitic foundation of the continent and shoving the entire mass eastward as a series of thrust sheets (b).

Based on a drawing by R.A. Price

syncline. Deep below the surface, the thrust fault along which the thrust sheet moved has carried these and older rocks upward and eastward onto the back of younger rocks. That is the characteristic signature of thrust faults in the Rockies: *older strata thrust upward and over younger strata.*

As you move eastward, shaly rocks of the Precambrian Miette Group are common in roadcuts. There you see more intense deformation. The rocks are tightly folded, with many small faults, and in places they have been subjected to heat and pressure sufficient to metamorphose the original shales into slates.

The mountains on the north side of the highway between Mt. Robson and Jasper are made of thick, hard quartzite of the Cambrian Gog Group. There the rocks form steep cliffs of gently to steeply dipping strata, each within a thick thrust sheet that has been moved eastward along a thrust fault such as the Pyramid Thrust at the base of Pyramid Mountain. Farther east you pass from one mountain range to another, from one thrust sheet to another. Each range, each sheet, is floored by a thrust fault. In places the sheets were bent during transport, such that the strata were folded into **anticlines** (up-folding) and synclines (down-folding). (See illustration in glossary, page 153.) In all of these ranges, the alignment of thrust faults and ranges is northwest/southeast, and most are inclined downward toward the southwest, in the direction in which the strata thicken and from which the sheets were moved. The Pyramid Thrust Fault, the Chetamon Thrust Fault, the Colin Thrust Fault, the Greenock Thrust, the Miette Thrust and the Brûlé Thrust all separate sheets or panels of strata that came from the west, where they once formed the ancient western shelf of North America.

WHAT OF THE FOOTHILLS?

Now you might ask, "What of the foothills? What about the Jurassic and Cretaceous strata that I see you have shown on the geological map? Do they have a story to tell?"

Indeed they do.

From the time North America rifted away from Australia until the first collision, a period of some 580 million years, all of the clastic sediment that accumulated upon the westerly sloping shelf came from the interior of the continent – from the *east*. By the beginning of the Paleozoic Era the mountain systems there had all but been eroded away. Thus, for the next 400 million years, only fine mud, silt and sand were carried westward by streams draining the interior and by marine currents circulating the waters of the shallow sea that covered the shelf. These fine sediments either reached the rim of the continent and were deposited in deep oceanic

Shale, siltstone and minor sandstone of the Cretaceous Brazeau Formation form some of the rocks of the Rocky Mountain Foothills east of Jasper National Park.

Photo by C.J. Yorath.

waters, or they found accumulation sites in low points on the shelf, such as between and beyond the reefs. So it was until the arrival of the Intermontane Superterrane. Then there were new sources of clastic sediment. This time from the *west*.

Imagine now that you are standing at the present site of Hinton, 75 km east of Jasper. It is about 170 million years ago, just after the Intermontane Superterrane has collided with the continent. Earthquake activity is intense, with large quakes occurring every few decades. You're patient. You stand for millions of years gazing far to the west, to where the edge of the continent meets the colliding edge of the Intermontane Superterrane. In the far distance the ground is rising; a surface bulge, like a gigantic ocean swell, begins to move toward you, much like a deep-water wave approaching a beach. In front of the advancing wave is a trough that moves with it, the two advancing together. Unlike an ocean wave, however, once the ground is uplifted it remains so; it doesn't subside as the wave advances. The moving bulge leaves in its wake the broadening highlands of the Canadian Cordillera.

Although the analogy with an ocean wave is not precise, it serves to illustrate the effects of the collision. Upon impact, the colliding edges were intensely deformed, metamorphosed and uplifted. As the squeeze continued, the entire sequence of rock layers deposited on the old continental

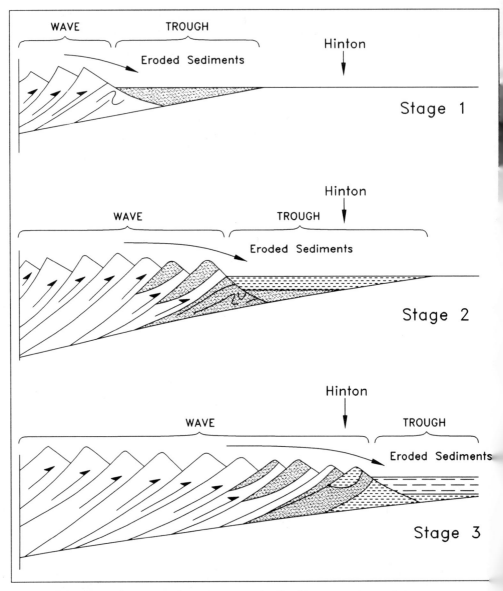

Like a gigantic eastward-moving wave, the Rockies were elevated along thrust faults. In front of the wave, a trough received detritus that had been eroded from the rising mountains (Stage 1). As thrusting continued, the eroded sediments in the trough also became incorporated into thrust sheets (Stages 2 and 3). This process continued until thrusting ceased about 60 million years ago, following which the mountain front has been eroded to its present position.

Based on a drawing from Gadd, *Handbook of the Canadian Rockies* (1995)

shelf came loose from its underlying Precambrian granitic foundation. Perhaps as much as 20 km thick, the entire pile began to move, breaking into thrust sheets, each a couple of kilometres thick. The moving thrust sheets, stacked one upon the back of the next, formed the growing bulge (the wave) that moved slowly eastward toward you, toward the interior of the continent.

In front of the advancing thrust sheets, a depression formed (the trough) as a result of the excess weight of rock that was loaded onto the crust. Clastic sediments, eroded from the uplifted impact zone as well as from the surface of the advancing thrust sheets, poured into the trough, themselves becoming part of newly rising thrust sheets as the deformation advanced eastward. This is the way in which the sedimentary strata and structures of the foothills were formed; the foothills consist of sediment worn from the growing mountains and dumped into the trough, then folded and thrust-faulted into mountains again, then eroded again into the trough. Recycling!

The Mesozoic sandstone and shale strata of the foothills, as well as the rocks of that age found in the front ranges of the Jasper Rockies, accumulated in three major episodes. The first two came in response to the two superterrane collisions with North America, while the third was a consequence of other types of deformational events that are not yet fully understood. Each event resulted in mountain-building. As they grew, the mountains eroded and spread sediments eastward into the trough at the foot of the ranges. Large deltas and **alluvial fans** built outward toward the centre of the trough, which was occupied by a narrow seaway that at times connected the Arctic Ocean with the Gulf of Mexico. The land was heavily forested, contributing woody and vegetal remains to the Cretaceous and Tertiary strata of the foothills, where today they occur as coal.

About 55 to 60 million years ago the pushing and shoving ceased. Although the total time it took to create the Cordillera was about 110 to 115 million years, it was primarily during the latter part of that interval that the Rockies were elevated. The enormous compressive stresses that had accumulated over millions of years relaxed, leaving the Canadian Cordillera in an unstable state. The result was that parts of the mountain system collapsed, the stresses that held them up having waned. By "collapsed" we mean that large blocks dropped downward along **normal faults,** steep northwest/southeast-aligned fractures separating blocks of crust, one of which has been dropped downward relative to the other (see illustration in the glossary, page 158). Throughout most of the Rockies it was the west-side blocks that dropped downward.

While there are several examples of normal faults in Jasper National Park, they are not as clearly evident as are the thrust faults. Thrust faults have clear topographic expression in that they separate whole mountain ranges. In our area the displacement on most normal faults is only a few metres, but in a few major normal faults the offset is many hundreds of metres. Normal faults in the Jasper area include the Cottonwood Fault, which passes through the town of Jasper, an unnamed fault passing across the lower parts of the ski runs on Marmot Mountain, and a complex array of normal faults on Mt. Fitzwilliam and surrounding peaks in the western part of the park. Others are described in Part Two of this book.

TEN KILOMETRES OF ROCK GONE DOWN THE RIVER

What goes up must come down, and on this planet we have a strong gravitational field to assist in turning mountains into plains. Consider all the agents of erosion that set to work on a mountain range the moment it rises above sea level. Waves pound at the shore; wind picks up beach sand and sandblasts the rocks; rain falls on the new land surface and gradually wears it away. Rivulets, streams and rivers carry rock fragments to the sea, at the same time carving their valleys ever deeper and wider. Masses of water-soaked soil slide down the slopes, and masses of rock break loose from cliffs, falling into the valleys below.

There are many other ways mountain ranges are eroded away: chemical weathering of rocks produces soil as freshly exposed minerals oxidize and dissolve; alternating expansion and contraction of rock surfaces through daytime heating and nighttime cooling loosens slabs and fragments; frost-shattering occurs when water enters cracks in rock and freezes, its consequent expansion forcing the cracks to open wider. Where a mountain range is high enough and a climate cool enough, glaciation wears away the peaks more quickly than any other erosional process.

For each mountain range in the Rockies, all of the rock that once connected the individual peaks has been eroded away. In other words, there has been more rock removed than remains today. The mountains you see are leftovers of the processes of erosion.

In the case of the Canadian Rockies, how much material has been eroded away?

Although there is no precise answer to this, it is possible to make a good guess. By measuring the amount of sediment being carried by Rocky Mountain rivers, we can get an idea of how much erosion is occurring at the moment. Geomorphologists (geologists who study the landscape) tell us that the Rockies are currently being worn down at an average rate of about six-hundredths of a millimetre per year. If that rate has been constant for the last 10 million years, about 600 m of rock have been removed in that time, a figure that agrees surprisingly well with estimates based upon the height of cavern entrances above the valley floors. A cavern system normally forms below the level of a valley floor, where limestone is saturated with water and dissolves, forming caves. Yet we find many cave entrances in the Rockies on the sides of cliffs 500–1000 m above the valley bottoms. These cave entrances have not been raised above the valleys, but rather the valleys have been cut deeper by streams and glaciers over a period of 6 to 12 million years.

As this last figure agrees with yet other independent measures, perhaps it is safe to conclude, conservatively, that the Rockies have lost between 600 m and 700 m of rock in the last 10 million years. Assuming a constant rate of erosion since mountain-building began about 120 million years ago, this gives us a minimum of about 6 km of rock removed, and perhaps as much as 12 km if we accept as good indicators those cave entrances currently 1000 m above the valley floors. If we split the difference, we can say that between 8 km and 10 km of rock have been removed by erosion over Jasper townsite.

Cave explorer Eric von Vorkampff-Laue stands silhouetted in the large entrance to Cliffside Cave, discovered in 1983 in the Snaring Mountain area northwest of Jasper. The entrance to the cave is 100 m below the top of a 670 m cliff; yet this cave, like others of its type, must have formed below the level of the valley floor. Millions of years of erosion have passed since then, carving the valley floor down to much lower elevations.

Photo by Ben Gadd

Does this mean that the Rockies were once eight to ten kilometres higher in elevation than they are now? No, it doesn't, and for two reasons.

The first is that erosion attacked the rocks as the mountains grew, thereby tending to reduce their elevation. The second reason is that because the earth's crust is depressed in proportion to the weight of additional load placed upon it, mountains on this planet cannot achieve an elevation of more than about 9 km above sea level. In the process of the Rockies' formation, the folding and thrusting of the rock layers had the effect of thickening the rocks on top of the underlying Precambrian granitic foundation. This increased thickness added to the weight or load on this foundation, which in response was depressed downward like a beam is depressed when a heavy load is placed upon it. Thus, the increase in the mass of rock forming the Rockies was compensated for by the sinking of the granitic foundations.

When the ranges were young and growing, the western part of the Rockies possibly approached the self-limiting height of the Andes or Himalayas. As time went by, the rate of mountain-building decreased until the rate of erosion exceeded that of upward growth. This would have lightened the load on the granitic foundation, causing the crust to rise back toward the level it occupied before the load was placed upon it. Thus, the processes of uplift and erosion are balanced by the condition of **isostasy:** the tendency of the crust, floating upon a weak, near-liquid underlying zone, to sink or rebound when weight is respectively added to or removed from it. In this way the Rockies have maintained much of their elevation, even though perhaps between 8 and 10 km of strata have been removed from above their summits. This process will continue until the Rockies ultimately disappear.

Where has all this material gone? Down the river, by and large. Jasper National Park covers about 11,000 km^2. Assuming that an average of 10 km of rock were removed from everywhere over this area, then 110,000 km^3 of mud, sand, gravel and dissolved limestone were disposed of. The park lies entirely within the region of Arctic drainage. All its surface waters drain into the Athabasca River, which, via the Slave and Mackenzie rivers, ultimately empties into the Arctic Ocean. The Mackenzie River Delta, a great pod of river-carried mud, silt and sand built out into the Beaufort Sea, contains material eroded from a vast area of the interior of the continent, including some of the 110,000 km^3 of rock eroded from the Rockies of Jasper. Another large portion of the material from these mountains accumulated to become the broad alluvial fans of the Tertiary Paskapoo Formation, which forms the surface strata of much of the foothills and adjacent plains of central southern Alberta. And the dissolved portion of the Jasper Rockies — the limestone portion, mainly — has added to the calcium content of the ocean.

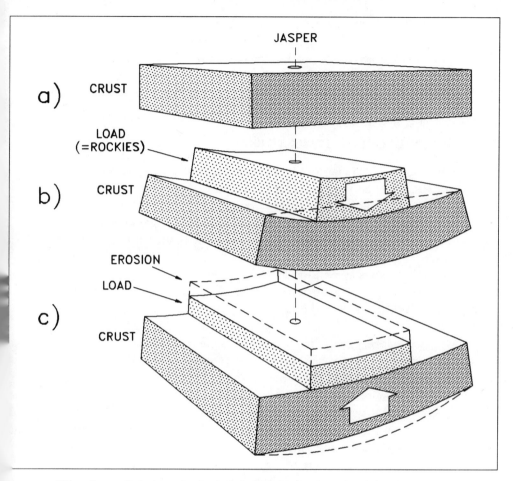

The effects of placing a load upon the earth's crust explain why the Rockies did not grow beyond a limiting height. Prior to the formation of the Rockies, the crust beneath the location of Jasper was a thick, undeformed slab floating upon a denser, liquid material *(a)*. When the Rockies (= load) were suddenly shoved eastward and placed upon the crust at the location of Jasper, the slab bent downwards under the load *(b)*. As the forces of erosion attacked the Rockies and wore them down *(c)*, the load on the crust was reduced, allowing the crust to begin to return to its former position. By this means the maximum height of the Rockies has been maintained at a more or less constant elevation over the past several million years, during which crustal rebound has kept pace with the rate of erosion.

We'll close this discussion by answering a final question you may be asking: "How long will the Rockies last?" That 0.06 mm/yr erosion-rate mentioned earlier can be worked forward as well as backward. The numbers suggest that in about 50 to 60 million years the remaining mountains will be gone, and the park will be reduced to a rolling plain much like the Canadian prairies. You had better get those pictures NOW!

ICE, THE GREAT SCULPTOR

The last chapter in the story of the formation of the Jasper Rockies has to do with glaciation, the most powerful eroder of all. One of the more obviously ice-carved peaks you can see from park highways is Mt. Edith Cavell, which is a fine example of an **arête,** a steep, narrow ridge whittled thin by glaciers carving away at both sides. Another is Mt. Fryatt, a **horn peak,** the result of glaciers gnawing from all sides to produce a horn-like central spire. Most peaks in the park display **cirques** on their flanks, where glaciers have quarried bowl-shaped amphitheatres, often cupping small, blue-green **tarn lakes;** at higher elevations many of these cirques still hold active ice.

It is entirely likely that glacial ice has periodically existed in the Canadian Rocky Mountains for millions of years, as is true for any high mountain range well north or south of the equator. However, it also is likely that the more recent glacial advances of the last quarter-million years have scraped away most of the evidence of these older glaciations.

Some evidence may remain, however, of one or more of these early glaciations. High atop peaks in the front ranges lie glacial **erratics:** pebbles, cobbles and boulders of quartzite carried by ice from the mountains to the west and left lying on glaciated limestone bedrock, far from the nearest quartzite outcrops. No one can yet say for sure when these erratics were put there, but it is quite possible that they were brought to their present position by an early glacial advance.

The last great continental and Cordilleran glaciation, and the one we know most about, is called the **Wisconsinan** glacial stage. From ice domes located far to the east and west, broad ice sheets grew outward, ultimately to coalesce into a single sheet of continental scale. Spreading westward from a centre in Hudson Bay, the **Laurentide Ice Sheet** reached the front of the Rockies, where it encountered tongues of ice extending eastward from the **Cordilleran Ice Sheet,** centred over British Columbia and covering the entire Cordillera to elevations up to 3000 m above sea level.

A wind-blown raven rests atop a pile of glacial till formed at the front of the Athabasca Glacier. Note the range in clast size, from large boulders (lower right) to small pebbles, all of which are enclosed within sand and silty clay.

Photo by Ben Gadd

A glacial erratic boulder of Cambrian Gog quartzite rests upon steeply inclined Devonian carbonate strata on Roche Miette.

Photo by Ben Gadd

At first the Rockies supported only valley glaciers, which moved like slow rivers down already established drainage courses, further carving out the broad Athabasca Valley as well as the valleys of the Astoria, Whirlpool, Maligne, Snake Indian, Rocky, Brazeau and many other rivers. Later, the Cordilleran Ice Sheet overwhelmed these valley glaciers, covering the ranges with so much ice that only the higher peaks stuck up above it as **nunataks.** Ice flowed eastward across the Rockies from British Columbia, carving deep, broad gaps such as Yellowhead Pass on Jasper National Park's western boundary. This eastward-flowing ice added its erosive power to that of the existing valley glaciers, which helps to explain the great width and depth of the Athabasca Valley, which carried the greatest flow of ice.

When did these glaciations occur? Recent studies have shown that, beginning about 2 million years ago, the earth's climate began a series of cold-warm cycles that caused the growth and decay of continental-scale ice sheets as many as eight times. The Wisconsinan glaciation began about 80,000 years ago, when the continental Laurentide Ice Sheet and the Cordilleran Ice Sheet intermittently expanded to reach their last maximum extents between 23,000 and 18,000 years ago. By about 10,000 years ago, the ice had largely melted and retreated from the Rockies and most of the remainder of the Canadian Cordillera.

We have one more glacial advance and retreat to deal with. Even though it was a minor event, it was the most recent. We are speaking of the **Little Ice Age,** known locally as the **Cavell advance** because, in Canada, it was first studied in detail at Mt. Edith Cavell. Beginning about 950 years ago, the world's average temperature declined and mountain glaciers began to expand. Glaciers grew larger throughout the park, in some cases extending a couple of kilometres farther down their valleys. They reached their farthest points of advance in the mid-1840s, after which the climate warmed and each glacier melted back, leaving behind its **terminal moraine:** a low ridge of till marking the farthest point reached. The ice was still fairly close to those moraines at the turn of the century, when the first non-Native explorers saw the Athabasca Glacier. Since the 1920s, people have been taking photos of this and other glaciers in the park; comparing these photos would show the degree of retreat throughout this century.

Will the retreat continue? Or will the glaciers advance again? Under natural circumstances, the present, relatively cool global temperatures should continue, thus permitting readvance and perhaps even the growth of continental ice sheets in the future. However, if average global temperatures rise as a result of increased atmospheric levels of carbon dioxide, a consequence of human activity (the greenhouse effect), then we may see continued ice retreat. At present there is intense debate among scientists as to which condition will prevail.

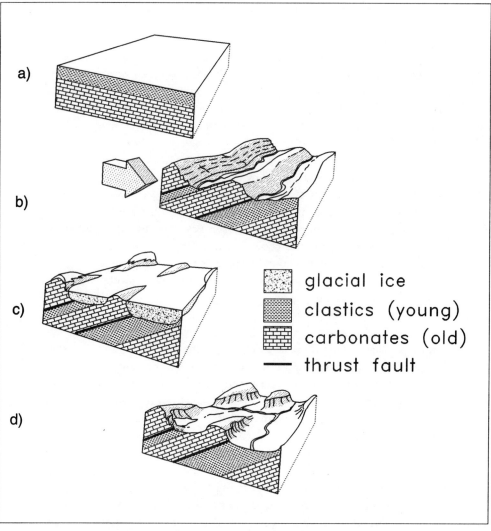

Development of the present physiography of the Rockies began with the accumulation of carbonate and clastic strata throughout some 580 million years (a). As a consequence of collisions between North America and far-travelled oceanic terranes, these strata were crumpled, broken into thrust sheets, and shoved northeastward. Erosion by rivers and wind cut into the thrust sheets, exposing tough carbonate layers forming the ridges as the softer shales eroded more deeply to form intervening valleys (b). During the past 2 million years glaciers have further eroded the landscape (c), leaving the shapes of the mountains and valleys we see today (d).

THE MOUNTAIN LANDSCAPE WE SEE TODAY

The 11,000 km² of mountains and valleys that make up Jasper National Park are bounded by several drainage divides that separate river systems flowing into the continent's surrounding oceans. The park's western boundary, which coincides with the provincial boundary between Alberta and British Columbia, also coincides with the **Continental Divide,** here separating the Pacific-bound flow of the Columbia and Fraser river systems from that of the Athabasca and Smoky river systems, whose waters eventually reach the Arctic Ocean. To the south, the boundary between Jasper and Banff national parks is defined by the drainage divide on Sunwapta Pass between the headwaters of the Sunwapta and North Saskatchewan rivers, the former a tributary of the Athabasca and the latter joining one of Canada's major rivers flowing into Hudson Bay. The eastern boundary of the park mainly follows the course of the Brazeau River (Hudson Bay drainage); north of that, the boundary follows an irregular course along the crest of the mountain front, connecting several minor divides as far as Mt. Lucifer at its northern apex.

As already noted, within the park, and indeed throughout all of the Canadian Rockies, most of the mountains are arranged in a series of parallel ranges aligned northwest to southeast, as shown on the geological map (in pocket). If you travel *parallel* to the ridges, as you do along the Icefields Parkway for example, physiographic and geological characteristics change little, but if you travel *across* them, such as along the Yellowhead Highway, these characteristics change dramatically from ridge to valley to ridge. The ridges are made of comparatively erosion-resistant rocks, such as carbonate and quartzite, whereas the intervening valleys have been cut in less-resistant strata, such as shale and siltstone.

In a general sense, the Rockies can be divided from east to west into three parts: the **foothills,** the **front ranges** and the **main ranges** (see page 36). Except for a small area in the southeastern corner, the foothills have been excluded from Jasper National Park. East of the park, however, the foothills occupy an area about 50 km wide. The rocks are mostly Mesozoic sandstone, shale, conglomerate and coal. They are soft and easily eroded, which explains why the foothills are not as high as the mountains.

The front ranges, as the name implies, form the mountain front immediately west of the foothills, and in Jasper National Park these ranges are about 30 km in width. As described previously, these ridges consist of thrust sheets of Middle Cambrian to Jurassic strata, each sheet having been thrust eastward over the one in front, such that the rocks above the faults are older than the strata beneath the thrusts (see the structural cross-section accompanying the geological map). In this way the ranges can be

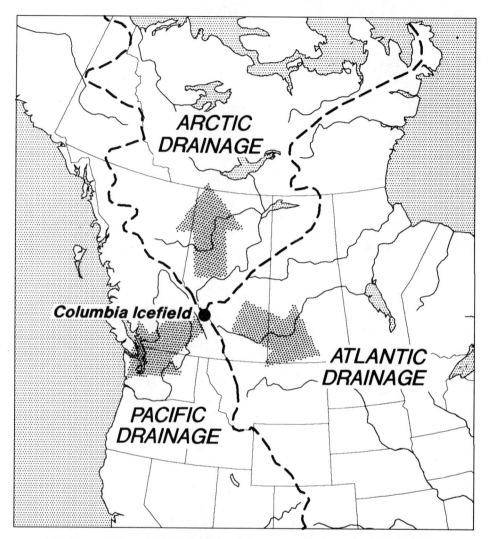

The two continental divides of North America, separating Pacific, Arctic and Atlantic drainage systems, intersect at the summit of the Snow Dome on the Columbia Icefield. This point is the hydrographic apex of North America.

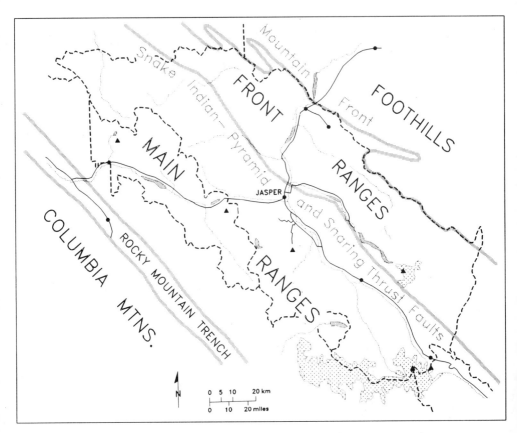

The Canadian Rockies are divided into main ranges, front ranges and foothills. In Jasper National Park and Mount Robson Provincial Park the main ranges consist of Precambrian gritstone and slate as well as lower Paleozoic carbonate and shale. The former strata commonly are intensely faulted and folded, whereas the latter consist of sheets of undeformed strata gently inclined to the west. The front ranges consist mainly of upper Paleozoic and Mesozoic carbonates and shales that occur as a series of westerly inclined thrust sheets and large folds. East of the front ranges the foothills consist of tightly folded and faulted Mesozoic and Cenozoic clastic strata.

thought of as a series of shingles, like those on the roof of a house. Each shingle-like thrust sheet is inclined to the southwest, and each has a steep, cliffy northeastern face and a gentle southwestern back slope. A good example is Chetamon Mountain, as seen to the north of the bridge over the Snaring River along Highway 16 east of Jasper. The steep, east-facing cliffs are made of erosion-resistant carbonates of the Cambrian Eldon and Lynx formations, whereas the intervening gentler slopes are composed of the more easily eroded shaly rocks of the Snake Indian, Pika and Arctomys (pronounced ARK-toe-miss) formations (see the geological time scale graph at front of book). The predominant colour is grey, the colour of limestone, which on some mountain sides has been bent into spectacular folds. It is grey cliffs such as these, reflecting a tinge of blue in the midday summer sun, that inspired Wilf Carter's nasal crooning of "In the Blue Canadian Rockies."

In contrast, the main ranges are much wider, spanning the Continental Divide (Jasper's western boundary) for a total width of 80 km. The main ranges of Jasper consist primarily of Cambrian and Precambrian strata of brownish and reddish hues. The thick Cambrian quartzite and carbonates forming peaks such as Mt. Edith Cavell and Pyramid Mountain are comparatively little deformed; they behaved as rigid masses during the collisional events that formed the Rockies. But the older, weaker Precambrian shale, slate and gritstone underlying the quartzite responded by folding and shearing. The difference is striking: strata of Cambrian quartzite forming the main cliffs of Mt. Edith Cavell, for example, are inclined gently downward toward the southwest, while the Precambrian slate beneath that cliff is wrinkled into many small folds and cut by numerous minor faults.

The main ranges can be divided further into eastern and western parts. Peaks in the eastern main ranges comprise mainly Cambrian quartzite and overlying carbonate strata that show little folding or faulting. The western main ranges, however, expose the underlying Precambrian gritstone and slate, which is intensely deformed. Moreover, the western main ranges show a higher degree of metamorphism, with garnet-bearing schists being common.

The main ranges are where the uplift of the Rockies was greatest. From the previous discussion of erosion and rebound, it should be no surprise that this also is where the greatest amount of rock has been removed, thus allowing for the exposure of the oldest, thickest and most deeply buried sedimentary rocks that accumulated across the rifted edge of the continent. In contrast to the main ranges, the front ranges consist of a thinner, younger succession of strata that accumulated upon the ancient shallow shelf of the continent.

The classic thrust-sheet style of the Rocky Mountain front ranges is shown by Chetamon Mountain, northwest of Highway 16 east of the Snaring River Bridge. The surface trace of the southwesterly inclined Chetamon Thrust Fault occurs at the base of the Snake Indian Formation, which forms the subdued slope beneath the prominent middle cliff of Eldon Formation carbonate, both of Cambrian age. Beneath the thrust fault the low peak to the northeast (right) is carved from Devonian and Carboniferous strata. Thus, by means of the thrust fault, older rocks of early Paleozoic age have been thrust over younger strata of middle and late Paleozoic age.

Photo by Ben Gadd

The boundary between the front ranges and main ranges consists of a system of thrust faults. From northern to southern Jasper park, the Snake Indian, Pyramid and Snaring thrusts form an overlapping system along which Cambrian and Precambrian strata, forming the main ranges, have been thrust eastward over Devonian and younger strata of the front ranges. Two and a half kilometres northeast of the town of Jasper, this boundary can be seen where Highway 16 crosses the **surface trace** of the Pyramid Thrust Fault (see page 109).

MORE ABOUT THE GEOLOGICAL MAP

As we described earlier, the various colours on the geological map that accompanies this book depict surface exposures of rock strata of specific ages, namely those of Precambrian (dark brown), Cambrian-Ordovician (dark blue), Devonian (purple), Carboniferous-Permian (mid-brown), Triassic (lime green), Jurassic (light brown), Cretaceous (forest green) and Tertiary (pink) ages. There are no rocks of Silurian age in this region. In very general terms the legend tells you the composition of these strata as well as something about the kinds of faults that disrupt them. The map portrays the ground surface as though shorn of its vegetation, soil and glacial debris, so that, were rocks of these ages actually coloured as they are on the map, the ground would indeed show this striped pattern when viewed from a high-flying aircraft.

The map has a striped pattern because the strata have been folded into anticlines and synclines and have been tilted, in most instances, downward to the west. The structural cross-section at the bottom of the map shows how these several layered systems of inclined strata crop out at the surface, where their bevelled edges form the striped pattern on the map, and how they have been brought into these positions by motions along southwesterly inclined thrust faults. On the map the heavy dark lines with triangles represent the surface traces of these thrust faults; the triangles are on the side, or thrust sheet, that was pushed upward and over the top of the strata on the other side of the fault. Lines with circles on them represent the surface traces of normal faults, where the side, or block, with circles on it has been dropped downward against the rocks on the other side.

The cross-section represents a vertical slice through the earth's crust drawn along a line extending from near Blackman Creek in the Selwyn Range at its southwestern end (SW) to a point about 15 km south of Hinton, Alberta, at its northeastern end (NE). Because we don't possess X-ray vision, the cross-section, and indeed the map as well, is an *interpretation* of the geological structure beneath the surface. The interpretation is reasonable because we have a good deal of information about the nature, distribution, thickness and geological relationships among the various rock units. This allows us to extend our surface observations into the subsurface.

Not shown on the map are the alignments of the many anticlines and synclines that occur within thrust sheets throughout the park. These were excluded to avoid unnecessary cluttering. In the front ranges, synclines can be identified where stripes of comparatively young rocks are flanked by stripes of older rocks; conversely, anticlines are seen where older rocks are flanked by younger strata.

In addition to the striped pattern, most particularly evident in the front ranges, there are other notable geological features shown on the map and cross-section. One of these is the length of many of the thrust faults. You also can see how these faults merge, one into another, both in the map view and in the cross-section. In the latter you can see how, according to

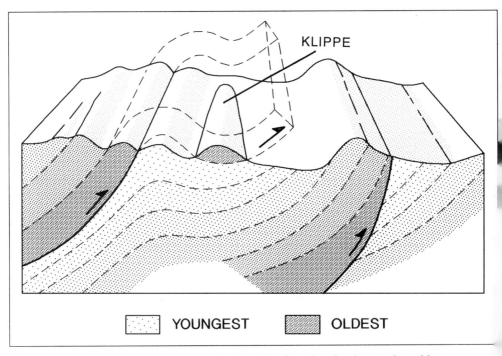

The classic signature of thrust faults in the Canadian Rockies is that older strata have been thrust upward and over younger strata. Together with anticlines and synclines these structures impart a striped pattern to geological maps, which show the distribution of rocks of various ages. This figure illustrates how a **klippe** (plural **klippen**) forms when erosion has removed much of the strata above a thrust fault, leaving only a remnant of a thrust sheet.

our interpretation, they ascend to the surface from the top of the Precambrian granitic foundation of the continent. At any given place along each thrust fault, the rocks above the fault are everywhere older than those immediately below the fault. This is a characteristic feature of all thrust faults in the Canadian Rockies. Moreover, it is thought that the thrust faults of the Rockies do not penetrate into the granitic basement, but instead flatten at depth and merge into a single surface of detachment, or **décollement,** at the base of the whole sedimentary pile.

Another interesting feature shown on the map occurs in the main ranges, where large regions of Cambrian rocks (ЄO) are separated by a comparatively narrow zone of Precambrian strata (PЄm) along the Miette River and Yellowhead Highway (Highway 16). Within this narrow zone, some of the older rocks of the Miette Group are exposed and several merging thrust faults show prominent deflections. Furthermore, where present, such as at Mt. Fitzwilliam, Cambrian strata appear as thin, isolated formations resting upon older Precambrian rocks. To the north and south the lower Paleozoic strata thicken substantially; at Mt. Robson, Cambrian strata are almost 8 km in thickness, and to the south, near Saskatchewan River Crossing in northern Banff National Park, these strata are more than 2 km thick. The implication of all this is that the narrow zone of Precambrian rocks along the Miette River represents a region where, during late Precambrian and early Paleozoic time, Miette Group and Gog Group sediments accumulated upon an east/west submarine ridge in the crystalline rocks forming the foundations of the continent.* Across the crest of this submarine ridge the accumulation of sediments was much thinner than to the north or south, a situation analogous to a headland extending seaward with deep bays on either side. The deflection in the orientation of thrust faults may be due to the presence of this ridge deep below the surface.

∽

Thus it was that our Jasper Rockies were formed. Over a period of some 750 million years the processes and forces that mould continents and oceans acted upon this region to create the magnificent scenery we enjoy today. Quartzite. Slate. Carbonate. Shale. Sandstone. Conglomerate. These are the materials of which the mountains are made. Rifting. Collisions. Folding. Thrusting. Erosion. Glaciation. These have all combined to shape the rocks into Mt. Robson, Mt. Edith Cavell, Roche Miette, Mt. Athabasca and hundreds more. Precambrian. Devonian. Carboniferous. Triassic. Jurassic. Cretaceous. Tertiary. These are the times when it all happened.

---

*You can see these Precambrian foundation rocks along Highway 5 south of Valemount, B.C., in the Malton Range.

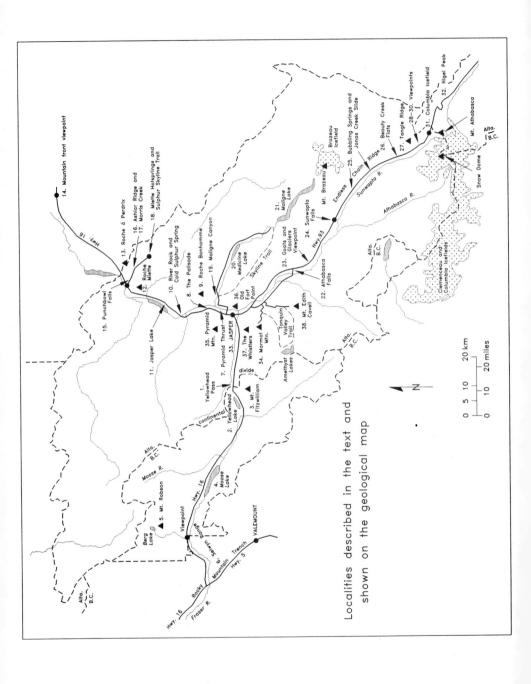

Localities described in the text and shown on the geological map

PART TWO

A GUIDE TO THE GEOLOGY OF JASPER NATIONAL PARK

In this part of the book we discuss the geology of many localities in the park. For the most part these places are readily accessible as viewpoints along the main highways, although some are accessible only by trail. We have concentrated on the main recreational destinations and other points of interest. For those localities along the main highways we provide distances from the intersection of the Yellowhead Highway (Highway 16) and the Icefields Parkway (Highway 93) in kilometres and miles. Localities along tributary roads are given from their points of intersection with Highway 16. Isolated points of interest off the main highways are also given, each with a description of its access route. On the accompanying geological map the locations of the main points of interest are numbered as they are in the text. Although not in Jasper National Park, Mt. Robson, in British Columbia's Mount Robson Provincial Park, is a part of the Rocky Mountains of the Jasper region. For this reason we include the highest mountain in the Canadian Rockies as a part of this guide.

The Yellowhead Highway (Highway 16)
Jasper to Tête Jaune Cache
Road distances given are from the intersection of Highway 16 and Highway 93 (Icefields Parkway) at Jasper.

Between Jasper and Tête Jaune Cache, roadside outcrops consist mainly of silty and limy shale, greyish-green, coarse-grained gritstone, and rusty brown-weathering slate *(6.1 km, 3.8 miles)*, all of the Precambrian Miette Group. The Miette Group's clastic sediments were the initial strata covering the torn edge of the continent following the separation of North America and Australia about 750 million years ago. These rocks, being comparatively weak, were more intensely deformed during mountain-building than were the thick, more rigid, or **competent,** Cambrian quartzite and carbonate formations forming the ridgeline seen north of the highway. The rest area at the east entrance to Mount Robson Provincial Park *(24.7 km, 15.3 miles)* displays some of these Precambrian strata folded into a small anticline (see photo on page 83).

For many kilometres west of Jasper, layers of grey-green gritstone alternate with layers of brown slate. Gritstone withstands erosion better than slate, so the gritstone layers tend to form low ridges on the valley floor, with the slate layers forming the hollows between the ridges. The repetition of gritstone and slate beds suggests that the rocks formed as a consequence of submarine landslides, perhaps caused by earthquakes. Evidence for this is found at the first roadcut west of the traffic lights at the intersection of Highways 16 and 93 *(0.6 km, 0.4 miles)*. Look closely at the gritstone exposed there; you will see that from the bottom of each steeply tilted bed, the size of the pebbles decreases upward (westward). At the top of each grit bed, the grains are very small, becoming so fine they form a layer of shale that has been metamorphosed to slate. These are called **graded beds.** Above each slate bed the sequence repeats. This phenomenon, together with other types of sedimentary structures, including **ripple marks** and **scour marks,** indicates that the repeated sets of strata are **turbidites.** The word comes from "turbid," meaning muddy, and each set of coarse-to-fine layers represents the deposit from a dense, underwater suspension of sediments moving rapidly, like a muddy landslide, down a steep submarine slope such as occurs at the edge of a continent.

Consider a single underwater landslide. The initial muddy rush of debris-loaded water scours the underlying seabed, ripping up soft pieces and carrying them along. As the strength of the current wanes, the sediment clasts begin to drop out. The larger, heavier fragments, including those ripped-up pieces, fall to the bottom first. Then the smaller, lighter ones

Rusty-weathering slates of the Precambrian Miette Group form many roadside outcrops along the Yellowhead Highway (Highway 16) west of Jasper. At this locality the stratification is nearly vertical, whereas planes of cleavage are inclined at about 45° to the right, forming the banded, cleaved surface of the outcrop.

Photo by W.R. Price.

settle out, in an orderly progression of decreasing size until, when the water is no longer moving, the tiny clay particles settle out of suspension and the water clears. An example of the size, power and speed of these currents was dramatically provided by the 1929 earthquake-induced submarine landslide on the continental slope of the Newfoundland Grand Banks. Several underwater communications cables, lying directly in the path of the landslide, were snapped by the passing dense current. By noting the times the cables were broken relative to when the earthquake occurred, Bruce Heezen and Maurice Ewing of the Lamont-Doherty Geological Observatory deduced that the current moved at an average speed of 45 km/hr over a total distance of more than 560 km!

Many of our clastic sedimentary formations were constructed this way, as turbidites, particularly those deposited in deep water. The greenish colour that you see in the gritstone beds of the Miette Group turbidites comes from the mineral **chlorite,** a compound of magnesium, iron, aluminum and silica formed by chemical alteration of other minerals. The white veins cutting through the rock in every direction are quartz, deposited in cracks in the rock by groundwater that was sufficiently hot, and under sufficient pressure, to carry quartz in solution.

Where did the sediment of the Miette gritstones and slates come from? As mentioned earlier, gritstone contains fragments of feldspar and mica in addition to grains of quartz, its most common constituent. These are the principal components of granitic rocks, so it is logical to infer that the source of the sediments was a granitic land mass. Quartz is tough and insoluble at natural surface temperatures and pressures, capable of surviving long trips down rivers to the sea, where it accumulates close to shore. Feldspar, on the other hand, is less resistant to weathering and erosion. Its chemical components are easily altered to clays. Thus, feldspar cannot survive long periods of weathering, nor does it hold up well during river transport. Mica is even more fragile than feldspar. The presence in Miette rocks of abundant coarse, angular clasts of quartz together with feldspars and mica shows that these sediments could not have been transported very far, nor could they have been long exposed to weathering. Thus, the source of the sediments was probably the granitic rock of the Precambrian Shield, and it could not have been very far away, perhaps less than 100 km.

The topography at the time of the landslide was quite rugged, as shown by the presence of pebbles in the Miette grits, for it takes a steeply sloping stream, with fast-moving water, to carry pebbles. A picture emerges of bare, lifeless mountain ranges standing close to the sea. Moaning winds, torrential rains and rushing rivers erode the ranges and carry their debris to the sea. Sediments accumulate, forming a narrow continental shelf bounded by a steep, unstable submarine slope. Here frequent earthquakes trigger landslides that cascade into deeper water to deposit their dense loads as turbidites, ultimately to form the Miette Group.

(1) Alberta/British Columbia Border: the Continental Divide
24.7 km, 15.3 miles

The elevation here is 1131 m above sea level, making Yellowhead Pass the second-lowest road or rail pass in the Rockies (at 869 m, Pine Pass, in the northern Rockies, is lower).

If you straddle the western boundary of Jasper National Park, you have one foot on the western slope and one foot on the northern – not the eastern – slope of the continent. For most of its course up the spine of North America, the Continental Divide separates streams draining to the Atlantic Ocean from those running to the Pacific Ocean, but not here. Jasper National Park lies just north of a second divide, this one coming in from the east and meeting the main continental divide at the Columbia Icefield. This second divide separates Alberta's watersheds into those that drain to the east and northeast through the Saskatchewan River system, Lake Winnipeg, and the Nelson River to Hudson Bay, which is considered to be Atlantic drainage, from those that drain northward via the Athabasca River to Lake Athabasca, the Slave River, Great Slave Lake, and finally the Mackenzie River, which empties into the Arctic Ocean.

(2) Yellowhead Lake
28.5 km, 17.7 miles

Yellowhead Lake, 32 m deep, is typical of water bodies found in Continental Divide passes of the Canadian Rockies. At the end of the last glaciation many of these comparatively low passes held stagnant ice, which became covered with sand and gravel carried by streams draining tributary valleys. The name for such material is **ice-contact stratified drift**. You can see some of these deposits in roadcuts along the lake. When the ice melted beneath this debris cover, the lake basin was formed.

The railway line beside Yellowhead Lake passes by high banks of ice-contact drift and lake silt. These unconsolidated deposits sometimes slide down onto the tracks, leaving railway maintenance workers the job of clearing them off.

(3) Mt. Fitzwilliam
36.8 km, 22.9 miles

The time to view the western face of Mt. Fitzwilliam is in the evening. In the light of the setting sun, you can peer back in time to when life on our planet suddenly exploded in numbers and diversity, to the end of the Precambrian Eon and the beginning of the Paleozoic Era (see photo on page 84).

Mt. Fitzwilliam, located just outside the western park boundary, is situated within a gentle syncline. The lower part of the mountain, reflecting a golden glow in the light of the setting sun, consists of shallow-water carbonates overlying silty shale, each a part of the Precambrian Miette Group. The upper part of the mountain consists of buff, pink and purple clastics of the Cambrian Gog Group, the bright colours hidden under a covering of grey lichens. The contact between the Precambrian carbonates and the

Yellowhead Lake, formed in a hollow left by the melting of stagnant glacial ice surrounded by gravel, is typical of many lakes in Continental Divide passes of the Rockies.

Photo by Ben Gadd

Cambrian Gog clastics is abrupt, representing a short gap in the geological record followed by a major change in environmental conditions that began between 590 and 570 million years ago. This change, resulting in deposition of fundamentally different types of sediments, coincided with the sudden increase in the diversity of life on earth.

Although life on our planet is thought to have begun as early as 3.85 billion years ago, it was not until 680–570 million years ago that cells with differing biological functions developed. This made it possible for many different species to exist, some with hard, preservable shell structures. By the beginning of the Cambrian Period of the Paleozoic Era, many different forms of life had developed shells and skeletal structures capable of being preserved as fossils. If, in fact, life was plentiful and diversified before this, we have little record of it; lacking hard parts, the organisms could not be preserved in rock. Occasionally, however, impressions of once-living Precambrian animals have been found, including those of jellyfish-like organisms, in the silty shales of Miette strata on the lower slopes of Mt. Fitzwilliam.

(4) Moose Lake
55.7 km, 46.8 miles

Moose Lake is 12 km long and 81 m deep. In part it may occupy a glacially deepened area of the Fraser River valley; however, deposits around the lake suggest a more complex origin. A hill of ice-contact stratified drift plugs the valley at its western outlet, forming a dam that would have pooled a lake regardless of the existence of an upstream bedrock basin. Following glacial retreat, a rockslide may have come to rest immediately upstream of the drift deposit, forming a slide heap large enough to create an additional dam that further raised the water level. Even later, a **debris flow,** consisting of mud and boulders, issued from a nearby creek bed and covered part of the rockslide.

(5) Mt. Robson
84.4 km, 52.4 miles

If you are travelling eastward on the Yellowhead Highway, the grandeur of Mt. Robson is clearly evident well before you reach the parking lot and viewpoint. When travelling west, however, it is suddenly and dramatically thrust upon you. Apart from being the highest mountain in the Canadian Rockies, with a summit elevation of 3954 m above sea level, Mt. Robson is also one of the more impressive mountains in North America (see photo on page 85). It has the greatest vertical relief (the difference between the

Moose Lake owes its origin to a complex interplay of several processes, including glaciation, rockslides and debris flows.

Photo by Ben Gadd

height of the mountain and the elevation of the adjacent valley floor), 3100 m, of any peak in the Rocky Mountains, even though many peaks in the Colorado Rockies are actually higher. The valley floor elevation at Mt. Robson is 853 m above sea level, whereas the highest peak in Colorado (4400 m) rises from an elevation of 2930 m, for an elevation gain of only 1470 m, less than half of that at Mt. Robson.

Mt. Robson is composed of Cambrian strata warped downward in a wide, gentle syncline, the same syncline that encloses the strata of Mt. Fitzwilliam. The mountain lies near the centre of the syncline, where the strata appear horizontal. The massive, greyish-brown lower cliffs are composed of carbonate strata of the Hota, Chetang, Tatei, Eldon and Pika formations. The buff-brown band about halfway up marks the Arctomys Formation, composed of red-and-green shale, siltstone and thin carbonate beds. Massive, cliff-forming carbonates of the Lynx Formation reach from the top of the buff band to the summit. In the lowest slopes, several outcrops of Gog Group quartzite occur along the trail to Kinney Lake near the base of the mountain. From there a spectacular, waterfall-decorated trail leads steeply upward to Berg Lake, commonly speckled with many small icebergs calved from the Tumbling Glacier (also called Berg Glacier), which plunges down the mountain's awesome north face.

On the flank of an adjacent peak, not far away, is Arctomys Cave, the deepest known cave in North America outside of Mexico. The cave has been formed in limestone of the Mural Formation, part of the Cambrian Gog Group. Two and a half kilometres long, Arctomys Cave descends steeply to a depth of 522 m below the surface. As a note of caution, the cave is cold and dangerous. Only experienced and properly equipped cave explorers should venture into it.

(6) Selwyn Range
100.0 km, 62.1 miles

Beyond Mt. Robson the highway passes around the northwestern end of the Selwyn Range, where roadside outcrops consist of gritstone and schist of the middle part of the Miette Group. Near the bridge over the Fraser River at Tête Jaune Cache, the gritstone beds have been intensely folded and stretched. If you are inclined to scramble among the low bluffs near the road, you will find places where these coarse clastic rocks have been acted upon by a deforming force powerful enough to stretch the quartz grains into long, parallel clasts, but not sufficient to affect the stronger feldspars, which resisted the stress, remaining as rounded grains.

The western boundary of the Canadian Rockies is defined by the **Rocky Mountain Trench,** part of a system of deep, linear valleys extending the entire length of the Canadian Cordillera and, according to NASA astronauts, clearly visible from the moon (see physiographic map on page xiv). Divided into northern and southern segments near Prince George, the Southern Rocky Mountain Trench is thought to be partly a fault-controlled valley, where the northeast side is a normal fault along which the valley floor was down-dropped. Farther north, around Prince George, the Southern Rocky Mountain Trench joins the Northern Rocky Mountain Trench, which, in turn, is aligned with the Tintina Trench extending through the Yukon. These latter two trenches were also developed along faults; however, unlike those of the southern trench, these faults separate parts of western Canada that moved horizontally past each other for distances of up to 750 km, between 100 million and 50 million years ago.

The Southern Rocky Mountain Trench separates the **Columbia Mountains** from the Rockies. On the western side of the trench lie the southern Cariboo Mountains and northern Monashee Mountains, two subdivisions of the Columbia Mountains. These ranges, like those of the Rockies, are very rugged, but the rocks are different, many of them being high-grade metamorphic rocks. The Columbia Mountains lack the orderly, shingled thrust-sheet structure and **physiography** of the Rockies.

Near the bridge over the Fraser River at Tête Jaune Cache, the low bluffs alongside the highway display what happens when rocks are stretched. Here, conglomerate consisting of coarse grains of quartz and feldspar was affected by forces sufficient to stretch out the quartz grains into long, streaky clasts – but not powerful enough to deform the stronger feldspar grains, which survived in their original shapes.

Photo by W.R. Price

The Yellowhead Highway (Highway 16)
Jasper to the Mountain Front
Road distances given are from the intersection of Highway 16 and Highway 93 (Icefields Parkway) at Jasper.

The drive eastward from the intersection of Highway 16 and the Icefields Parkway takes you past several thrust sheets of Upper Paleozoic carbonate rocks. Each sheet is separated from the others by thrust faults, along which the rock strata broke when the mountains began to form. As a result of a massive shove from the west, the thick succession of Proterozoic to Triassic strata that makes up the Canadian Rockies was broken into separate layers by these faults, such that each successive sheet was thrust (pushed) upward and eastward onto the back of the one forming in front of it. By this means the ranges of the Rockies were stacked progressively higher while simultaneously being eroded by rivers and later carved by glaciers into the mountains and valleys we see today.

The nature of the rocks in these thrust sheets has had a profound effect on the landscape. Each thrust sheet represents a tilted slab of rocks resulting from thrust-breaking of the same overall set of formations; hence, from slab to slab, or range to range, the same formations are repeated – older over younger, older over younger. Thus, as you pass through the front ranges, you see the same set of formations again and again. That set includes mainly tough limestone in the lower part and relatively weaker siltstone and shale in the upper part of each tilted slab. Because siltstone and shale are more easily eroded than limestone, after some millions of years the more resistant carbonates now remain as prominent ridges while the less resistant clastic rocks underlie the valleys. Because thrust-faulting has repeated the rock sequence six times between Jasper and the mountain front, what you see are six limestone ridges, each one a separate range, and six intervening siltstone-and-shale-floored valleys. It's a classic example of **structural control** of a landscape, where the underlying geology dictates the physiography.

A serendipitous set of coincidences related to latitude and climate also contributes to the grand shapes of peaks in the front ranges. A typical front-range peak has a gentle southwestern slope and a steep, cliffy northeastern slope, primarily because of the southwesterly inclination of the strata. The southwest side of a mountain anywhere in Canada is the warmest side, because at this latitude the southwest side receives the most sunlight late in the day, when air temperatures are highest. Thus, the southwest side of a peak is likely to have the least snow accumulation in winter, and the snow there melts early in the summer. In contrast the northeast slope is the coldest. In winter it receives little sunlight (the sun is

The physiography of the front ranges, with their carbonate ridges and shale-floored valleys, is well illustrated by this aerial view of the valley of Vine Creek.

Photo by Ben Gadd

low on the southern horizon), and what little it gets comes in the morning, when the air temperature is low. Therefore, snow falling on the northeast slope is likely to stay through the winter and last well into the summer. Add to this the direction of the wind in this part of the world – mainly from the west. This wind tends to blow snow up the gentle southwesterly side of a typical front-range peak and over the ridgecrest onto the northeast side, where it further contributes to accumulation. The net effect of these circumstances is the development of glaciers on the northeast side of our peaks, glaciers that quarry cirques into mountain slopes. As a consequence, the northeast sides of front-range ridges are commonly serrated with cirques, bounded by cliffs hundreds of metres in height.

During the ice ages the valleys between the ridges also held glaciers. These cut into the valley walls, undercutting the lower ends of the sloping limestone slabs found on the southwestern sides of most front-range ridges. When the glaciers melted, the slabs slid down into the valleys, leaving smoothly sloping surfaces from valley to ridgeline such as you see on most front-range peaks (see illustration on page 105).

Because the limestone ridges and valleys are aligned in a northwest-to-southeast direction, the layout of stream courses in the front ranges takes on a regular, parallel pattern. The ridges on either side have small streams draining their slopes, and these meet the valley trunk rivers at approximate right angles. This pattern, called **trellis drainage,** is characteristic of the front ranges throughout the Canadian Rockies.

The classic amphitheatre-like shape of this cirque on Roche Boule is characteristic of many such features on the eastern slopes of the Rockies.

Photo by Ben Gadd

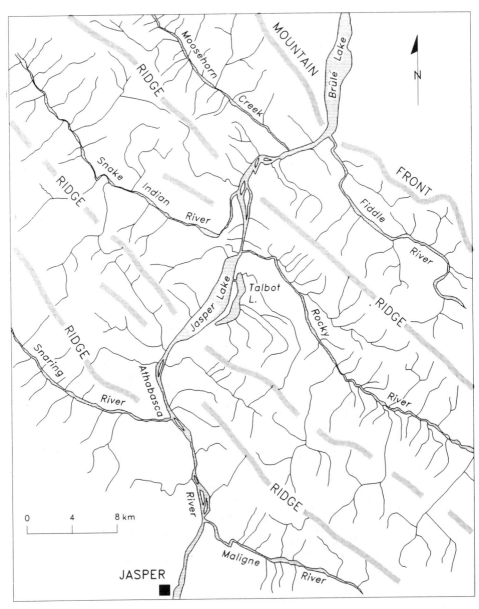

The pattern of trellis drainage comes from the development of antecedent and subsequent streams. Antecedent streams such as the Athabasca River existed before mountain-building and were able to maintain their courses across the ranges as the mountains rose. Subsequent streams such as Moosehorn Creek, Fiddle River, Snake Indian River, Rocky River, Snaring River and Maligne River developed during mountain-building, and their courses were set by the alignment of the folds and faults.

What, then, of the Athabasca River valley, which cuts across all the high limestone ridges?

To explain this, we need to call upon the concept of **antecedent streams.** A look at a map of Alberta shows that most of the rivers flowing away from the Rockies flow northeastward. Within the Rockies, the major rivers also flow northeastward. It's unlikely that these rivers started out by flowing southwesterly or northwesterly along the valleys of the front ranges and then somehow managed to cut across the intervening ridges. Instead, it is much more likely that these rivers have been flowing to the northeast ever since the Rockies began to rise. A powerful river such as the Athabasca was able to maintain its course through the growing stack of thrust sheets, cutting downward as quickly as the ridges rose across its course. That is what antecedent drainage is: stream courses formed *before* the mountains through which they flow were built.

Smaller streams, on the other hand, could not keep up with mountain-building. They were forced to follow the lay of the land, with its parallel ridges and valleys, producing the trellis drainage discussed above. Such streams are termed **subsequent,** because they formed during and *after* mountain-building.

On with our tour.

(7) Boundary between the main ranges and front ranges
6.8 km, 4.2 miles

The surface trace of the Pyramid Thrust Fault crosses the eastern shoulder of Pyramid Mountain where it descends to the level of Highway 16, about 1 km north of its junction with the Maligne Road (metal bridge across the Athabasca River). There the fault is inclined to the southwest at about 45°. At this locality, Pyramid Creek follows the surface trace of the fault, which separates the main ranges from the front ranges in this part of the Rockies. Here Precambrian rocks of the Miette Group have been thrust over Devonian carbonate strata of the Flume Formation and black shale of the Perdrix Formation, each well exposed at road level. The Perdrix shale looks like coal, but the black colour is from the carbon-rich remains of sea life, not plants.

This is a good place to compare the shapes and colours of mountains in the two belts. Looking to the northeast along the highway, you see steeply inclined slabs of Paleozoic rock displaying typical front-range colours: grey limestone and brown shale. Looking back toward Jasper, you see rounded reddish/brownish main-range mountains of Precambrian gritstone and slate, with the great Cambrian quartzite prow of Mt. Edith Cavell off to the southwest. Looking south, across the Athabasca River, you can

The surface trace of the Pyramid Thrust Fault follows the treed gully in the left foreground and forms the boundary between the front ranges and main ranges. The low, rounded hill is in the front ranges and is formed from Devonian carbonate (pale grey) and black shale, respectively of the Flume and Perdrix formations.

Photo by C.J. Yorath

see a low cliff just above the trees, at the base of the mountain. That cliff exposes Precambrian carbonate rock and clastics that lie on the main-range side of the Pyramid Thrust. Near the east side of the outcrop you can see Carboniferous carbonates that lie on the front-range side of the fault; they are separated by a **splay** (additional small fault) of the Pyramid Thrust.

(8) The Palisade
8.3 km, 5.2 miles

A "palisade" is a type of fortification, a term befitting the prominent northeast-facing cliffs on the west side of Highway 16, opposite the Palisade picnic area. Several Paleozoic formations are exposed in the cliffs, and although the strata appear to be flat-lying along its face, the layers actually dip gently away into the cliff toward the southwest, where they are overlain by Cambrian and Precambrian rocks in the Pyramid Thrust Sheet.

At the base of the cliffs are carbonate rocks of the Cambrian Lynx Formation, the same formation exposed at the top of Mt. Robson. The Lynx is overlain by interbedded limestone and shale of the Upper Cambrian and Lower Ordovician Survey Peak Formation. The remainder of the cliff consists of formations of Devonian and Carboniferous age. (See upper photo on page 86.)

Did you notice that the sequence here jumps from strata of Ordovician age to rocks of Devonian age? What about those of the Silurian Period? As discussed earlier, there are no rocks of Silurian age in this part of the Canadian Rockies. In a sequence, the absence of strata of any age marks what is called an **unconformity**. The explanation for this particular unconformity is that, before the latter part of the Devonian Period (prior to 375 million years ago), the continental shelf was elevated above sea level. Any Silurian strata deposited before that time were removed by erosion (if indeed they had been deposited in the first place). In fact, in this part of the ancient shelf, represented by strata of the Palisade, the unconformity represents the absence of strata of most of the Ordovician Period, all of the Silurian Period, and all of the early and middle parts of the Devonian Period. In other words, 85 million years of the earth's history have left no record in this region!

Above the unconformity, which coincides with the contact between the Ordovician Survey Peak and Devonian Flume formations, the remaining part of the cliff reveals black shale of the Perdrix Formation, richly fossiliferous silty and sandy shale of the Mount Hawk Formation, sandstone and siltstone of the Sassenach Formation, and the thick, massive limestone of the Palliser Formation, which makes up the upper half of the cliff. Above the Palliser, and just out of sight over the top of the cliff, are interbedded limestone and shale of the Carboniferous Banff Formation.

(9) Folds and faults in the front ranges
12.4 km, 7.7 miles

As you pass by the Jasper airstrip at what the locals call "Henry House Flats," look back over your shoulder to see the northeast side of Roche Bonhomme and adjacent Grisette Mountain. There the entire Cambrian to Triassic succession has been bent into a huge syncline above the Chetamon Thrust Fault. As you continue northeastward along the broad Athabasca Valley, past marshes and lakes, you cross several other thrust faults, each separating panels of folded Paleozoic, Triassic, Jurassic and Cretaceous strata. Just before reaching Pocahontas and the road to Miette Hotsprings *(44.5 km, 27.6 miles)*, you get a magnificent view of the castellated peak of Roche Miette (see photo on page 88). A short distance

Looking southeast from Henry House Flats, you can see how the folded carbonate and clastic strata of Grisette Mountain (left foreground) and Roche Bonhomme (centre far distance) together form a large syncline. As you move your gaze from Grisette Mountain to Roche Bonhomme, your eyes follow an inverted arc from downwardly inclined carbonate strata of the former to steeply upwardly bent strata in the latter.

Photo by W.R. Price

farther on, near the bridge over the Fiddle River *(50.0 km, 31.1 miles)*, you can see a low mountain to the north (Bedson Ridge) in which Devonian and Carboniferous rocks have been folded into an anticline (up-fold) that is lying on its side (lower photo on page 86). Just before reaching the eastern park gate, you will see the imposing Roche à Perdrix.

(10) River Rock and Cold Sulphur Spring
21.9 km, 13.6 miles

On the far side of the highway bridge over the Athabasca River, stop to take a look at the smoothly sloping grey cliff beside the road. This is known locally as "River Rock." The Athabasca River strikes it full on, and if you watch the river for a while, you'll see a cycle of strongly upwelling currents every few minutes as the water piles up, then escapes downstream.

The rock is Devonian limestone of the Palliser Formation, one of the principal cliff-forming formations of the Canadian Rockies. The weathered

A glacially smoothed surface of Devonian Palliser limestone forms the steep face of River Rock above the Athabasca River.

Photo by Ben Gadd

surface of River Rock is light grey, though a freshly broken piece shows that the rock is dark coloured. The black colour is due to the presence of small quantities of hydrocarbons (organic material) distributed evenly through the rock, whereas the mineral calcite, which is what a massive limestone like the Palliser Formation is mostly made of, is white. Hydrocarbons that are exposed to the air oxidize: that is, they join with oxygen and break down into carbon dioxide and water. It is reasonable, then, to expect that the black hydrocarbons on a fresh surface of Palliser limestone will ultimately disappear, leaving only the white calcite crystals showing. You can get an idea of the speed of this process by looking at the north end of River Rock, which was broken off in the early 1950s when the highway was built. The black Palliser surfaces exposed at that time are about halfway to going grey.

A walk across the bluffs overlooking the river shows that the Palliser Formation contains easily recognizable fossils, including brachiopods and corals. Look for prominent glacial grooves carved into the limestone.

The large parking area just east of River Rock allows you to stop and approach Cold Sulphur Spring, a stream of water issuing from the base of the cliff. The rock is Devonian Flume Formation limestone. The water smells of dissolved hydrogen sulphide gas, just like Miette Hotsprings or the hot springs near Banff. The spring here is not hot (although at 9°C it's about four degrees warmer than the soil temperature), but the presence of the gas shows that the water originates several kilometres below the surface, within an environment similar to that of other hot springs throughout

The rotten-egg smell of hydrogen sulphide gas causes Cia Gadd to hold her nose beside Cold Sulphur Spring, which issues from beneath a cliff of vertically layered Devonian limestone of the Flume Formation.

Photo by Ben Gadd

the Rockies (see page 77). Cold Sulphur Spring is cold because the hot water has mixed with normal groundwater on the way up to the surface.

The strange grey material in the bed of the stream running from the spring is not a mineral deposit. It's alive! These are colonies of harmless bacteria that live by extracting energy from sulphur.

One last bit of geology invites your attention here. If you stroll east along the outcrop from which Cold Sulphur Spring flows, you will come to an example of the sub-Devonian unconformity, the proper name for the

The rock record of some 135 million years of earth history is missing along the sub-Devonian unconformity, here represented by a thin layer of orange and yellow pyrite (at person's hand) separating Cambrian Lynx Group dolomite (left) from dark limestone of the Devonian Flume Formation (right).

Photo by Ben Gadd

gap in the geological record described on page 59. It's not easy to pick out, because the beds above and below the unconformity are tilted at the same angle and look much the same, but here is a clue: beside a tree growing at the base of the rock, there is a place where the rock goes from grey Upper Cambrian Lynx Group dolomite, with no fossils, to black Upper Devonian Flume Formation dolomite, with white rice-like fossils of the stromatoporoid *Amphipora*. The exact erosion surface representing the unconformity is marked by a thin zone of rusty-weathering pyrite.

(11) Jasper Lake
29.3 km, 18.2 miles

If you visit Jasper Lake in the summer, it will probably be there. If you come in winter, however, it won't. This is because the lake fills in the early summer and drains in the late fall.

Jasper Lake is extremely shallow. If you don't mind wading in cold, glacier-fed water, you can get about halfway out into the lake before the water reaches much above your knees. At about that point you will probably find one of the frequently changing channels of the Athabasca River.

This is not a lake, it is simply a moving sheet of very shallow water that develops annually with spring flooding. At its eastern end two rivers join the Athabasca: the Rocky River from the southeast and the Snake Indian River from the northwest. Each has built a broad alluvial fan (fan-shaped, sand-and-gravel deposit) across the valley floor, and the Athabasca River is thus obliged to flow between them. Each spring, snowmelt upstream increases the flow of the Athabasca, which backs up behind these alluvial fans, spreading out over the valley floor and creating Jasper Lake. Glacial meltwater keeps the flow strong through midsummer, but in the fall, when the melt rate declines, the Athabasca slips back into its winter channels and the lake dries up (see upper photo on page 87).

The speed of water flowing through the lake is fast enough to carry fine, clay-sized particles in suspension but too slow to carry sand and silt-sized particles; these are left behind. Every year, with each new flood, a new layer of sand and silt is deposited on the lake bed, and every winter, when the lake is dry, the strong winds that blow from the west down this valley pick up the silt and sand, producing dust storms that sweep out into the foothills. Sand particles are too heavy to be carried far, airborne, by wind; instead, they move by rolling or bouncing along the ground. At certain locations in the valley where the strength of the wind sharply declines, the sand and silt grains accumulate in dunes. The silt has been deposited separately as **loess** (pronounced *luss*), seen in prominent light-coloured patches farther east along the highway.

The stillness of a summer evening is evident in distant reflections in Jasper Lake, which seasonally fills as a consequence of spring snowmelt upstream along the Athabasca Valley (top). In the fall, when glacial meltwaters diminish, the lakebed dries, exposing silt and sand to winter winds (bottom). The consequent dust storms sweep out into the foothills. The sand piles up in dunes.

Photos by Ben Gadd.

The Rocky River deposits its sediment load and braids its channel prior to joining the flow of the Athabasca River.

Photo by Ben Gadd

(12) Roche Miette
41.3 km, 25.7 miles

Seen from beyond the park's east gate, the battlements of Roche Miette appear as a massive wall of limestone, not unlike the famous El Capitan cliff in the Guadalupe Mountains of west Texas. As you pass through the gate and drive about 10.2 km (6.3 miles) westward, the mountain appears end-on, as illustrated in the photograph on page 88.

Roche Miette is one of the more striking mountains in the Rockies. It is situated in the axis of a syncline formed above the steeply inclined Miette Thrust Fault, the surface trace of which occurs on the low, rounded slope northeast of the peak. Above the fault are Cambrian and Devonian strata that form the syncline, the former exposed along the low ridge to the northeast and the latter exposed as nearly horizontal formations in the walls of the main peak. The lower slopes of the main peak are composed of dark shales of the Perdrix Formation, successively overlain by shaly limestone of the Mount Hawk Formation, clastic strata of the thin Sassenach Formation, and the massive limestone of the Palliser Formation, which forms the impressive upper cliff.

(13) Roche à Perdrix
50.0 km, 31.1 miles

Roche à Perdrix, at the east gate to the park, is structurally more complicated than Roche Miette. The Devonian Perdrix and Mount Hawk formations have been deformed into a complex series of anticlines and synclines beneath massive carbonates of the Palliser Formation that form the top of the mountain (see photo on page 89). Within the Palliser, small thrust faults rise upward out of the core of a syncline, a common characteristic of folded carbonates.

(14) The mountain front
55.1 km, 34.2 miles

The eastern boundary of Jasper National Park follows the crest of the **mountain front,** an impressive wall of grey rock over a kilometre high. From the southern end of the Canadian Rockies to north of Jasper, the mountain front displays this same marked contrast: rugged cliffs of limestone stand high above the rounded slopes of the foothills, which are made of sandstone and shale. Why the abrupt change in rock type at the mountain front? And why the big difference in elevation?

Seen from the highway east of the park, the mountain front appears as a wall-like carbonate ridge rising above the subdued topography of the foothills.

Photo by C.J. Yorath

Let's consider the rock type first. Originally the sandstone and shale beds of the foothills were laid down over the limestone beds seen in the front ranges. They were all part of the same flat-lying stack of layers, with limestone and other carbonate rocks in the lower part and sandstone and other clastic rocks in the upper part. During mountain-building, the stack broke, and the lower part was thrust-faulted upward at a low angle toward the northeast, over the upper part. Reaching the surface, the limestone layers slid along on top of the sandstone and shale layers as a massive, thick thrust sheet. The leading edge of this thrust sheet moved as much as 40 km before the push from the southwest died out and the sliding stopped. The point reached by the thrust sheet was, by definition, the mountain front. Northeast of the mountain front lay more sandstone and shale, while the mountain front itself was made of the limestone of the thrust sheet. That is why the change in rock type is so abrupt: it lies along a fault.

The difference in elevation between the foothills and the front ranges is largely the work of erosion. That may sound strange, because it is easy to imagine a thick thrust sheet hulking over the land surface on which it moves, forming a mountain range in itself. But keep two things in mind: (1) erosion was going on throughout mountain-building, and (2) those thrust sheets still carried the sandstone and shale of the upper part of the stack.

More than a mountain's height of rock was worn away from the front ranges *as they formed,* so the difference in elevation between the thrust sheets and the land northeast of them might not have been all that great – until the upper part of the stack was gone, exposing the tough limestone layers underneath. When that happened, erosion of the front ranges slowed down. Erosion of the soft foothills, though, continued as rapidly as ever. Over time, the foothills lost more elevation than the front ranges did.

Regardless of the actual height of the mountain-front thrust sheet, the front ranges were destined to stand taller simply because the rock in them was harder. It's the same process that produced the limestone ridges and shale valleys within the front ranges: differential erosion. In the 60 million years that have passed since both the foothills and the front ranges formed, differential erosion has resulted in an elevation difference of 1200 m between the two.

The Hoff Anticline is typical of many large folds in the front ranges of the Canadian Rockies.

Photo by Ben Gadd

The Fiddle Road
Punchbowl Falls to Miette Hotsprings and the Sulphur Skyline Trail
Road distances given are from the intersection of Fiddle Road and Highway 16.

Visitors travelling to Miette Hotsprings pass by much that is of geological interest: evidence that the Rockies were born as a desert mountain range, the awesome limestone wall of Ashlar Ridge, and the important geological message contained in the twisted layers of the Fernie Formation. Once at the hot springs, consider hiking the trail to the Sulphur Skyline summit. It's well worth the effort.

(15) Punchbowl Falls
1.3 km, 0.8 miles

Mountain Creek flows over a steeply tilted bed of hard conglomerate of the Cretaceous Cadomin Formation to form Punchbowl Falls, a waterfall. If you follow the short trail down to the lower viewpoint, you can see how the water has eroded a slot into the formation, with a lovely plunge pool at the bottom.

The Cadomin conglomerate is a thin but widespread formation in the foothills and front ranges. It contains pebbles of quartzite from the Gog Group and **chert** (a mineral made of microscopic quartz crystals) from the Rundle Group, the sources of both having been to the west in the main ranges. The formation is Early Cretaceous in age, indicating that the Gog and Rundle rocks from which the pebbles were derived were above sea level and being eroded about 115 million years ago.

The Cadomin Formation tells us something about the climate of the Rockies at the time they began to rise. The rock looks much like the thin sheets of gravel that overlie bedrock surfaces at the bases of desert mountain ranges. Desert ranges erode in such a way as to produce an abrupt change in slope at the base of the mountain front. The gravel-covered desert floor rises gently upward to meet the mountain front, from which point the relief rapidly steepens toward the peaks. In such places erosion occurs in pulses: periodic rainstorms cause flash floods capable of moving large quantities of sand and gravel, and these sediments accumulate as broad alluvial fans at the mountain front. The Cadomin Formation possibly represents such a mode of sediment transport and accumulation, and its thinness over such a large area is further evidence of its origin on a **pediment** – the bedrock surface over which the pulses of gravel move and are deposited (see illustration on page 72). Thus, this part of the Rockies may have been born in a desert climate.

The lovely waterfall at Punchbowl Falls spills over conglomerate strata of the Cretaceous Cadomin Formation.

Parks Canada photo

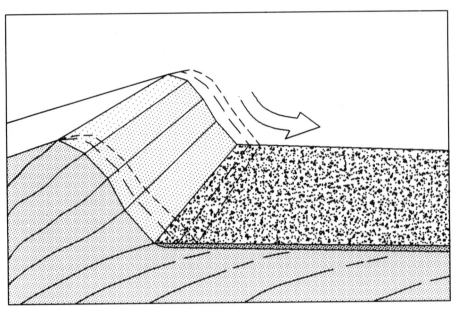

A pediment is a low-relief bedrock surface over which gravels eroded from adjacent mountains are deposited as thin sheets during episodes of flash floods. Pediments develop in desert environments.

Overlying the Cadomin conglomerate are layers of sandstone and shale of the Luscar Group which, at the lower viewpoint, contain a thin coal seam. From 1911 to 1921, 800,000 tonnes of coal were mined from a much thicker seam nearby at the Pocahontas Mine, the entrance to which is not far from Punchbowl Falls. Long abandoned, the mine has been sealed, and no further mining is permitted in Jasper National Park.

⑯ Ashlar Ridge
8.6 km, 5.3 miles

After passing through forest for several kilometres, the Fiddle Road turns a corner and suddenly presents a dramatic view of a spectacular mountain wall: Ashlar Ridge (see lower photo on page 87). This is one of the more impressive cliffs in the front ranges of the Rockies, with a summit elevation of about 1830 m above sea level. The sheer face is 300 m high, made of grey-weathering Devonian Palliser Formation limestone strata that are almost vertically inclined.

Ashlar Ridge is yet another product of differential erosion. A close look at the gullies leading up to the base of the cliff reveals the presence of softer layers – shales of the Carboniferous Banff Formation and carbonates

At Punchbowl Falls, coal is found within Cretaceous Luscar Group strata. At the nearby Pocahontas Mine, some 800,000 tonnes of coal were mined between 1911 and 1921.

Photo by Ben Gadd

of the Rundle Group — overlying the harder Palliser Formation. As these strata are less resistant to erosion than the tough Palliser limestone, they have been worn away to a greater extent, leaving the Palliser Formation to form the ridge. Over the last 2 million years, glaciers flowing down the Fiddle River valley have added to this erosion by scraping away the softer layers stacked against the Palliser ridge.

(17) Fernie Formation at Morris Creek
9.0 km, 5.6 miles

On page 53 we described how the underlying geology has set the pattern for the physiography of the front ranges: carbonate ridges and shale valleys. It's easy to see the bedrock in the limestone ridges, but the valleys are covered with vegetation and good exposures of the shale are rare. Here at Morris Creek and at nearby roadside outcrops to the southeast, one can get a look at the Jurassic Fernie Formation, the softest of the valley-floor formations.

Shale is a clastic rock made of tiny particles of mud. Coarser, heavier particles — such as sands and pebbles — when carried to the sea are commonly deposited in shallow waters close to shore. Muds, on the other hand, accumulate in deeper, quieter waters where the fine, lighter particles settle out of suspension; these, after being buried beneath further accumulations, harden into shale.

The Fernie Formation is the last sequence of sedimentary materials to have been derived from erosion of older rocks to the east. Until about 150 million years ago, all clastic sediments that accumulated upon the western shelf of the continent came from the east, from the interior of the continent. All of that changed with the collision of the Intermontane Superterrane and the ancient continental edge. Then there were new sources of sediment, this time from the west. Beginning about 170 million years ago, the colliding edges were moulded into the uplifted **Omineca Belt** (the Selwyn,* Cassiar, Omineca, Cariboo, Monashee, Selkirk and Purcell mountains), which thenceforth provided the sediments that were transported eastward and deposited as alluvial fans and deltas in front of the rising mountain system.

Some strata of the Fernie Formation contain **belemnites**, cigar-shaped fossils from an extinct group of squid-like animals (cephalopods) that had a hard internal shell, rather like a cuttlebone. Other beds contain

---

* Of the Yukon, not to be confused with the Selwyn Range of the Rockies mentioned on page 51.

At Morris Creek, Jurassic Fernie Formation shale and siltstone strata have been folded into a small anticline that has been truncated by a minor fault.

Photo by Ben Gadd

abundant clams and **ammonites,** which were coiled sea-going cephalopods, progenitors of the modern nautilus.

Like other shales in the Canadian Rockies, those of the Fernie Formation were highly deformed during mountain-building. The strata are so twisted and broken that it is difficult to follow them for any distance. Fortunately, however, thin layers of orange-weathering siltstone stand out boldly, providing a sense of form to these complex structures.

(18) Miette Hotsprings
17.0 km, 10.6 miles

Open from Victoria Day to Labour Day, Miette Hotsprings is one of Jasper National Park's main attractions. Few people who enjoy them, however, know where the water comes from or how it is heated.

Contrary to some published explanations, the water flowing from the springs has not simply trickled down through the mountain and flowed out at its base, somehow heating up along the way. For the water to acquire these temperatures, it must have been heated within the earth at a depth of at least 3 km, where the rock temperature is at or above the boiling point of water. Rock at such depths is normally saturated with water, and at these pressures water remains liquid at higher temperatures than it does at the surface. It also readily dissolves sulphur minerals, which is how it acquires the rotten-egg smell of hydrogen sulphide gas characteristic of most of the Rockies' hot springs. The presence of the gas is another indicator that the water has come from well below the surface.

In most circumstances, hot groundwater is unable to return

The pool at Miette Hotsprings is fed by groundwater that is heated at a depth of over 3 km before returning to the surface along the Hotsprings Thrust Fault.

Photo by Ben Gadd

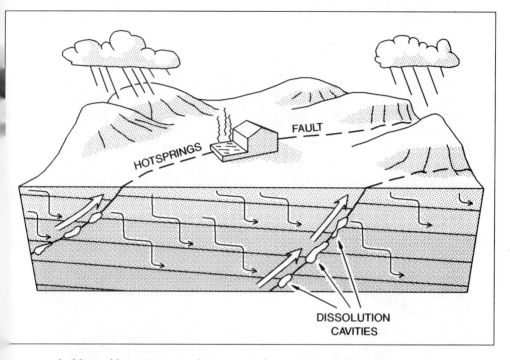

At Miette Hotsprings, circulating groundwater is heated to high temperatures at depths of 3 km or more before it returns to the surface along the Hotsprings Thrust. The hot water dissolves limestone along the fault, widening its conduit and thus allowing more water to reach the surface before it loses all of its acquired heat.

to the surface quickly enough to retain its heat. This is because its movement through layered rocks is very slow. To retain a portion of its heat, it must move quickly to the surface through a natural conduit. Natural conduits are provided by fractures such as thrust faults, particularly those involving carbonates, which on dissolution further enhance the speed of the water's travel to the surface. Most hot springs in the Rockies are found close to thrust faults, which suggests that they use these faults as conduits, the water rising upward because it is hotter and thus less dense than the surrounding fluids. Miette Hotsprings, for example, flows from the appropriately named Hot Springs Fault.

As you walk along the paved path beyond the old pool site to a hotsprings outlet (one of three here; the others are enclosed), you will pass large boulders of **tufa,** which consist of spongy-looking masses of porous, friable (easily crumbled) calcium carbonate and gypsum encrusting angular blocks of limestone, small and large. Dissolved carbon dioxide and hydro-

One of three outflows of Miette Hotsprings water has a temperature of 51°C.

Photo by Ben Gadd

Deposits of spongy-looking, porous, friable calcium carbonate and gypsum, called tufa, are common along the path to the old pool site at Miette Hotsprings.

Photo by Ben Gadd

gen sulphide in the hot-springs water escape as the water cools, causing the precipitation of calcium compounds from solution. In this case, the tufa has developed close to the stream and on adjacent slopes where limestone rubble has become incorporated into tufa deposited by springs that no longer flow. At the modern outflow you can see bright-yellow sulphur, formed as the dissolved hydrogen sulphide gas reacts with atmospheric oxygen when the hot-springs water reaches the surface. The temperature of the water at the outflow is 51°C.

On your return to the modern pool you may notice several boulders of fossiliferous dolomite in the stream bed. These are derived from the Devonian Cairn Formation, which is exposed not far away in a carbonate reef built by stromatoporoids; the growth of this reef is illustrated on page 14. Beneath the plains of central Alberta these reefs are called the Leduc

Formation, which forms the reservoirs for much of the province's oil and gas.

Sulphur Skyline Trail
4 km one way, elevation gain 700 m

The Sulphur Skyline Trail takes you above the timberline, where the views are truly spectacular. Although long and steep, the ascent to the summit of Sulphur Ridge, elevation 2070 m, can be accomplished without special mountain-climbing equipment.

From the Miette Hotsprings parking lot you will find the trailhead to the right of the main entrance to the pool building. It begins as a wide dirt track used by maintenance vehicles, but soon narrows to a fine walking trail that climbs up a narrow valley underlain by Carboniferous carbonates and shales. A forest fire cleared much of the timber off this slope many years ago, so views are unimpeded.

The trail reaches a pass at km 2.2 and branches. Keep right. At the timberline, it passes by a large glacial erratic of white Gog Group quartzite around which spalled chips litter the ground. The bedrock along the ridge leading to the summit is Triassic siltstone of the Sulphur Mountain Formation. At the summit is limestone of the Carboniferous Mount Head

A glacial erratic boulder of Gog Group quartzite lies along the trail leading to the summit of Sulphur Ridge above Miette Hotsprings.

Photo by Ben Gadd

Formation, the uppermost formation of the Rundle Group. All of these strata have been folded into an anticline, with the summit of the mountain located at its crest. The west limb of the anticline is truncated by the Hot Springs Fault. Following the ridge westward and down from the summit for about 100 m brings you to the surface trace of this fault, where the Livingstone Formation, the lowest formation in the Rundle Group, has been thrust over rocks of the Mount Head Formation.

The view from the summit is superb. To the northwest you can see the full length of Ashlar Ridge, a cliff of Devonian Palliser limestone 6 km long and 300 m high (see lower photo on page 87). To the southwest you can look across the valley of Sulphur Creek to the cliffs of Utopia Mountain, which expose a Devonian reef of the Miette reef complex (see page 14). To the southeast the land descends 700 m to the valley of the Fiddle River, winding between rugged limestone peaks of the Nikanassin Range. To the northeast you can see over the top of the mountain front to the green forests and rolling topography of the foothills.

On the way back down the trail, as your knees are telling you that they've had enough, keep thinking about the pleasant soak awaiting you in the hot pool.

View from the top of the Sulphur Skyline. The shale-floored valleys of the Fiddle (right foreground) and Moosehorn (left background) rivers are flanked by carbonate ridges, one of which is Ashlar Ridge in the right mid-ground.

Photo by Ben Gadd

THE MALIGNE VALLEY
Road distances given are from the intersection of Highway 16 and the Maligne Lake Road.

The following is a round-trip tour. The first part, from Maligne Canyon to Maligne Lake, is by road; the second, from Maligne Lake back to the canyon, is by foot and involves a two- to three-day backpack along the magnificent Skyline Trail.

Of all the road-accessible valleys in Jasper National Park, the valley of the Maligne River is the most geologically varied. Lying along the surface trace of the Pyramid Thrust Fault, the Maligne Valley separates the front ranges to the northeast from the main ranges to the southwest. Views are outstanding. A deep limestone gorge, Maligne Canyon, lies just off the highway, offering a safe, well-designed interpretive trail. A cave system lies below the valley floor, carrying the Maligne River underground for part of its length. The valley holds the longest natural lake in the Canadian Rockies and the second-largest known rockslide. It also contains Medicine Lake, which looks like a reservoir but isn't. These and other attractions bring hundreds of tour buses into the valley each year.

(19) Maligne Canyon
5.9 km, 3.7 miles

An apocryphal story has it that in the early 1950s, while visiting Maligne Canyon during her stay in Jasper for the filming of the non-epic *River of No Return*, Marilyn Monroe fell into the canyon and was heroically rescued by Robert Mitchum. It is almost certain that nobody told the actress, who was said to have injured her ankle in the process, that the rock through which the canyon is carved is limestone of the Devonian Palliser Formation.

The canyon is situated just east of the surface trace of the Pyramid Thrust Fault, which forms the boundary between the front ranges and main ranges. The former are composed mainly of Devonian and younger, dominantly carbonate strata, and the latter mainly of Cambrian and Precambrian clastic rocks, with fewer carbonate strata. Along the thrust fault, coarse-grained quartz sandstone and gritstone of the Precambrian Miette Group, resistant to mechanical and chemical weathering, have been thrust over the more easily dissolved Paleozoic carbonates into which the canyon has been cut.

There are two possible ways to explain the formation of Maligne Canyon. The process may have begun following retreat of glacial ice in the Maligne Valley about 12,000 years ago. The northwesterly flowing Maligne Valley glacier met the thick, northeasterly moving *(continued on page 99)*

Water roils through the gorge of Maligne Canyon throughout the summer, making it a very dangerous place to explore, but in the early spring the flow is only a trickle, enticing these park visitors to clamber over the water-polished rock surfaces and the potholes on the floor of the gorge.

Photo by Ben Gadd

At the east entrance to Mount Robson Provincial Park, Precambrian Miette Group strata have been folded into a small anticline, the east flank of which has been eroded away.

Photo by W.R. Price

Against the setting sun, Precambrian carbonates of the Miette Group reflect a pale yellow glow beneath the dark purple tones of the Cambrian Gog Group in Mt. Fitzwilliam on the western boundary of Jasper National Park.

Photo by W.R. Price

With a peak elevation of 3954 m, Mt. Robson is the highest and one of the more impressive mountains in the Canadian Rockies. Composed mainly of Cambrian carbonates, the mountain is the centrepiece of Mount Robson Provincial Park in British Columbia.

Photo by Ben Gadd

The carbonate wall of the Palisade includes strata of Cambrian, Ordovician, Devonian and Carboniferous ages, spanning an interval of some 160 million years of earth history. Devonian Palliser Formation limestone forms the steepest part of the cliff.

Photo by C.J. Yorath

On Bedson Ridge, Devonian and Carboniferous strata of the Palliser and Banff formations respectively have been folded into an anticline that is lying on its side.

Photo by Ben Gadd

From the air the winter channels of the Athabasca River are clearly visible beneath the shallow waters of Jasper Lake.

Photo by C.J. Yorath

The impressive wall of Ashlar Ridge is composed of steeply inclined Devonian strata of the Palliser Formation.

Photo by C.J. Yorath

The castellated peak of Roche Miette near the east entrance of the park is carved from nearly horizontal Devonian carbonate strata of the Palliser Formation. The cliffy middle slope is formed from shaly carbonates of the Mount Hawk Formation, and the lower slopes from easily weathered shaly rocks of the Perdrix Formation. On the ridge to the left, steeply inclined beds of the Cambrian Eldon, Pika and Arctomys formations occur beside Devonian strata of the Flume, Maligne and Perdrix formations.

Photo by W.R. Price

Thick, massive carbonates of the Palliser Formation form the main peak of Roche à Perdrix. Palliser limestone may be seen at road level along Highway 16 at the east gate of the park.

Photo by W.R. Price

Medicine Lake, dammed by a landslide at its northern end (prominent hill at the end of the lake in the middle distance), is located above an underground karst drainage system. In spring and summer meltwater inflow exceeds the capacity of the drainage system, and thus the lake fills (top). In fall and winter, inflow is greatly reduced and the lake drains.

Parks Canada photos

In the middle distance at the foot of the prominent cliff, a rock glacier of jumbled talus has moved downslope from an amphitheatre-like basin at its head. The imposing peak in the background is the Watchtower, consisting of Precambrian and Cambrian strata dislocated by a complex series of stacked thrust faults.

Parks Canada photo

Against the early-morning sun, the Cambrian quartzite cliffs of Mt. Tekarra look like a fortress constructed upon a hill. The hill consists of Precambrian Miette Group strata.

Photo by C.J. Yorath

Cambrian strata of **Mt. Kerkeslin** are warped into a gentle syncline between the **Simpson Pass Thrust** on the southwest (right) and a normal fault to the northeast (left).

Photo by W.R. Price

Along the Icefields Parkway the highway traverses the path of a landslide near Jonas Creek. Clearly visible high on the ridge is the scar marking the spot from which the pink rocks of Cambrian Gog Group quartz sandstone slid down into the valley.

Photo by W.R. Price

The deeply crevassed Athabasca Glacier, as seen from Icefield Centre on the Icefields Parkway, is one of several glaciers draining the Columbia Icefield.

Photo by Ben Gadd

At Athabasca Falls the Athabasca River carves its course through Cambrian quartzite strata of the Gog Group.

Photo by C.J. Yorath

As though wearing a death mask, the seemingly sightless eyes of the snow-covered face of "The Old Man" (Roche Bonhomme) stare upward into a blue-grey sky above the town of Jasper. The upper part of the face is composed of Triassic siltstone of the Sulphur Mountain Formation, resting upon Permian and Carboniferous strata.

Photo by Ben Gadd

Across the valley from the Stutfield Viewpoint on the Icefields Parkway (Highway 93), the Stutfield Glacier, part of the Columbia Icefield, flows over a vertical cliff of Cambrian limestone. The resulting ice avalanches send plumes of ice and snow cascading into the valley below.

Photo by Ben Gadd

The impressive triangular mass of Pyramid Mountain appears reflected in the still, early-morning waters of Pyramid Lake above the town of Jasper. The upper part of the peak is divided from the lower by the Snaring Thrust, forming a klippe. The entire mountain is formed from northeasterly inclined Cambrian clastic strata of the Gog Group.

Photo by W.R. Price

Named in honour of the executed British nurse who remained behind to treat wounded soldiers after the fall of Brussels to the German army in World War I, Mt. Edith Cavell, south of Jasper, is thought by many to be the queen of the Canadian Rockies. Gently inclined Cambrian quartzite forms its steep scarps, which are separated by thin layers of snow-covered sandstone and siltstone.

Photo by W.R. Price

With powder streaming in their wake, two skiers traverse the slopes of Marmot Mountain, unaware of a 750-million-year-old story of continental rifting recorded in the Precambrian rocks lying beneath the snow. Across the Athabasca Valley stand the buttressed Cambrian quartzite cliffs of Mt. Tekarra.

Photo by Hugh Lecky, courtesy of Brian Rode, Ski Marmot Basin

Cambrian quartzite of the Gog Group forms the battlements of the Ramparts above the Amethyst Lakes in the Tonquin Valley.

Photo by Ben Gadd

Athabasca Valley glacier, which, at the point of intersection, had carved the Athabasca Valley more than 120 m deeper than the Maligne Valley. Upon retreat of the ice, the Maligne Valley was left **hanging,** that is, with its floor higher than that of the Athabasca Valley. The Maligne River would thus have cascaded into the Athabasca Valley, perhaps as a waterfall. Subsequent to ice retreat, the river above this point became confined to its present course on the north side of a lateral moraine, and was thus forced to flow across Palliser limestone rather than through the nearby shaly Carboniferous strata that form the lower slope beneath the Pyramid Thrust Fault. The result of these processes was the creation of a gorge cut by the waterfall as it eroded its course upstream.

Alternatively, it has been suggested that the canyon may have begun as a cave system, the roof of which was breached by glacial erosion. In this scenario, the Maligne Valley glacier would have carved through the overlying shaly rock, exposing the cave beneath. In many places the width of the canyon is greater at the bottom than at the top, with many dome-like chambers typical of Rockies caves. It is probable that the canyon existed before the end of the last ice age, because glacially derived sediment remains in the canyon (from First Bridge, look for boulders in the large pocket high up on the canyon wall). Further, it is notable that the Maligne River is much reduced in this part of the valley, and even dry during the winter, making it debatable that the river could have cut such a deep canyon in only 12,000 years. Most of the flow is underground, in a cave system connected to the canyon at many points; if you follow the trail down the canyon past Fourth Bridge, you can see water flowing from a cave entrance. Further evidence for the unroofed-cave hypothesis is offered by the existence of a similar canyon upstream, one branch of which goes into a large cave (known as Mousehole Cave), as well as by the presence of small stalactites in Maligne Canyon.

Canyon formation is the result of two complementary processes, mechanical weathering and chemical weathering. The mechanical weathering is caused by the abrasive action of sediment-laden water passing over bedrock, which is gradually worn away. The chemical weathering can be attributed to the solubility of limestone in water, here made slightly acidic by the presence of dissolved carbon dioxide.* Hence chemical weathering plays an important role at Maligne Canyon, as it does elsewhere throughout the Rockies, where limestones host a variety of caves, underground rivers like the Maligne, and several other forms of **karst** topography.

---

*Additional acidity comes from weak solutions of sulphuric acid arising from the presence of pyrite in the rocks, a compound of iron and sulphur.

The interior of Mousehole Cave, a limestone cave in a gorge along the Maligne River, is illuminated by light from the "Mousehole." This short cave is all that remains of a larger cavern unroofed by glacial erosion; the unroofed portion is the gorge. Maligne Canyon, a few kilometres downstream, may have had a similar origin.

Photo by Ben Gadd

Regardless of its mode of origin, the form of Maligne Canyon is controlled by the presence of **joints** in the limestone. Joints are fractures in the rock on either side of which no appreciable motion has occurred. It is the presence of joints, together with stratigraphic layering, that gives many outcrops their angular, blocky appearance. In the Maligne Valley there are three perpendicular sets of joints. One is parallel to the strata, which are inclined gently toward the southwest at about 10°. The other two sets are nearly vertical. One is aligned northwest/southeast, more or less along the canyon. The other is oriented northeast/southwest, at right angles to the first, which accounts for the sharp, short bends in the canyon's course.

In the upper part of the canyon, near the teahouse, you can find **potholes**: rounded and smoothed depressions worn into the bedrock by the grinding action of gravel, cobbles and even boulders trapped in low spots in the canyon floor and swirled round and round by the water. When the water level is low, you can see the stones in the bottoms of these potholes. On a larger scale, the curved, smoothed walls of the canyon show where fast-moving water preferentially dissolved and mechanically eroded away the rocks on the *outsides* of curves. The concave walls on *both* sides of the canyon are typical of dissolved cave passages, but different from the walls of stream-cut canyons, which tend to form parallel curves. In a few places, very large boulders are caught between the narrow walls of the canyon, forming natural bridges.

Park naturalist Cleone Todgham emerges from a deep pothole in Maligne Canyon. Potholes form when water-swirled pebbles drill their way down into the rock.

Photo by Ben Gadd

(20) Maligne River and Medicine Lake

The valley of the Maligne River between Maligne Canyon and the southeast end of Medicine Lake *(27.4 km, 17.0 miles)* roughly coincides with the boundary between the front ranges and main ranges of the Rockies. The front ranges, lying to the northeast of the Maligne Road, consist of Paleozoic carbonate rocks in a shingled series of southwesterly inclined thrust sheets. The Colin Range forms this side of the valley and consists mainly of Devonian carbonate strata of the Palliser Formation. The mountainside is formed along the plane of inclination, or **dip,** of the strata. To the southwest, Cambrian and Precambrian strata of the Maligne Range have been thrust eastward over Paleozoic strata along the Pyramid and Snaring thrust faults; together these faults form the boundary between the main and front ranges. A hike up Watchtower Valley *(trailhead at 17.5 km, 10.9 miles)* to the northwest brings you into beautiful alpine country. Above the valley the imposing guardian mass of The Watchtower exposes Cambrian and Precambrian strata in a complex series of stacked thrust faults and **klippen** (isolated parts of a thrust sheet that remain when the rest has been eroded away; see illustration on page 40).

For the most part, Medicine Lake and the outflowing Maligne River lie on limestone of the Devonian Palliser Formation. The lake formed after a huge slab of Palliser limestone slid down into the valley from the dip-slope surface on the southwest flank of the Colin Range, thus damming the flow of the Maligne River. The volume of the slide has been estimated at 86,000,000 m^3, making it one of the larger slides in the Jasper Rockies.

The lake itself is shallow, seasonally filling and emptying depending upon the relative balance between inflow and outflow, the latter occurring underground for some 15 km through the slide rubble and by means of caves within the Palliser limestone (see photos on page 90). In late summer, when inflow from the Maligne River is low, the underground drainage system can carry away all the water in the lake, leaving a dry lake bed throughout the fall and winter. During these seasons the river braids its way across mudflats to sinkholes in the limestone located along the northwest and northeast shores of the lake. In June, when run-off from snowmelt increases, inflow greatly exceeds the capacity of the sinkholes to carry water away. Thus, the lake fills, sometimes overtopping a low point in the slide. At such times, about once every five years, a surface stream flows out of the lake; otherwise the lake has no surface outlet.

Where does the water go? By adding red rhodamine dye at the sinkholes and watching springs that were thought to issue Maligne Lake water, researchers have learned that the water resurfaces in and about Maligne Canyon and elsewhere in the Athabasca Valley, including the floor of Lac

Beauvert in front of Jasper Park Lodge. In summer the water takes about a day to make the trip from Medicine Lake to Maligne Canyon; in winter, when the flow is greatly reduced, it takes about a week. Based on the straight-line distance between Medicine Lake and Maligne Canyon, this underground karst drainage system may be the longest in Canada. We offer a prize of one chocolate bar to anyone who finds an unblocked, explorable entrance to the system and returns to tell us about it.

Southeastward from Medicine Lake, the road runs largely through rockslide debris as it traverses Carboniferous, Permian and Triassic strata. It then crosses into a klippe of Cambrian Gog Group sandstone thrust over brown-weathering Triassic siltstones and sandstones of the Sulphur Mountain Formation, the latter surrounding the northwestern end of Maligne Lake. Along this route, striking evidence of massive rockslides is provided by cabin-size boulders beside the highway. Some of these are enormous masses of chert that tumbled into the valley from the Permian Ranger Canyon Formation, which outcrops high in the Queen Elizabeth Range, the northeast wall of the valley. Carbonate rockslide boulders litter the Maligne River at *34.7 km (21.6 miles)*. The largest is known as "Rosemary's Rock." Closer to the lake, rockslide debris consists of quartzite from the Cambrian Gog Group, and siltstone from the Triassic Sulphur Mountain Formation.

This huge boulder of chert from the Permian Ranger Canyon Formation tumbled into the Maligne Valley during one of several landslides that profoundly modified the landscape of this region.

Photo by Ben Gadd

If you have the impression that rockslide debris litters the floor of the Maligne Valley from Medicine Lake to the slide damming Maligne Lake, you are right. What accounts for so many slides? Dave Cruden, a University of Alberta professor, has studied many rockslides throughout the Canadian Rockies. He has found that slides are most common in places where carbonate strata are inclined at about 30°–40° *and* where the slope of the mountainside is a little steeper than the inclination of the rock strata. This is precisely the situation in the Maligne Valley. It seems most likely that glacial erosion oversteepened the northeastern wall of the valley, and when the ice melted away, the unsupported slabs slid down into the valley bottom.

Rosemary's Rock and other boulders that lie in the Maligne River provide further evidence of landslides in the Maligne Valley.

Photo by Ben Gadd

(21) **Maligne Lake**
41.6 km, 25.8 miles

The "jewel of the Canadian Rockies," as many like to call it, Maligne Lake owes its origin to both glaciation and a landslide. The southern half of the lake is very deep (97 m), suggesting that this part was excavated by moving ice, or perhaps dammed behind a moraine. In contrast, the northern half is notably shallower, suggesting that the original glacial lake did not extend this far down the valley until a huge rockslide dammed the outflowing Maligne River, resulting in expansion of the lake to today's dimensions.

At an estimated volume of 498,000,000 m^3, this slide is the second-largest known in the Canadian Rockies.* It originated in strata of Carboniferous, Permian and Triassic age from the "Sinking Ship," a local

---

*The largest is in the Valley of the Rocks near Mt. Assiniboine – one billion cubic metres.

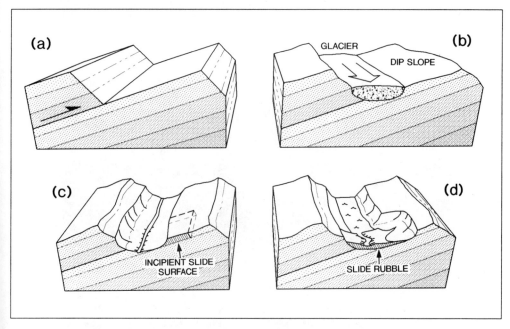

Landslides are common in the Maligne and Athabasca valleys and occur where glacial erosion has oversteepened dip slopes underlain by carbonate or quartzite strata. A glacier erodes a valley *(a)*, producing a steeper bedrock valley wall slope than the inclination, or dip, of the stratification *(b)*. Subsequent frost action or other mechanism loosens the now-unsupported slabs, which descend into the valley as large landslides *(c)*.

name given to the peak that overlooks the concession area and parking lots at the northwest end of the lake. Another slide, this one originating in the Opal Hills – the low, brownish mountains northwest of the Sinking Ship – also contributed to the natural dam. As you approach the lake along the Maligne Road, you climb a hill made of brown siltstone from the Opal Hills Slide. Nearer to the shore, boulders of carbonate from the Sinking Ship Slide may be seen, and a view down the lake reveals house-sized carbonate blocks on the right (southwest) shore, all part of the same enormous slide heap. Only recently have the rockslide origins of Maligne Lake been recognized; formerly the lake was thought to have been dammed by a glacial moraine.

The surface trace of the Pyramid Thrust, which in this part of the Rockies defines the boundary between the main ranges and front ranges, follows the southwest shore of Maligne Lake, then turns away from the lake and continues along the course of the Maligne River, which enters the lake about halfway along its southwestern shore.

The glacier-mantled peaks of Mt. Brazeau and Mt. Mary Vaux at the southeast end of Maligne Lake are carved from carbonate and clastic strata of Cambrian to Devonian age. The lake owes its origin both to overdeepening of the valley floor by glaciation and to the damming of the Maligne River by the second-largest known landslide in the Canadian Rockies.

Photo by Ben Gadd

The debris and hummocky topography of two landslides are ▶ visible from Opal Peak near Maligne Lake. The Sinking Ship Slide (top), with an estimated volume of almost 500,000,000 m², is the second-largest known slide in the Canadian Rockies and was partly responsible for the formation of Maligne Lake. The Opal Hills Slide (bottom) was smaller, but nonetheless brought down a mass of reddish-brown Sulphur Mountain Formation siltstone large enough to reach from the slide heaps shown in the photo to the valley floor.

Photos by Ben Gadd

MALIGNE LAKE / 107

Maligne Lake, 22.3 km long and 97 m deep, is the longest natural lake in the Canadian Rockies. The tour-boat cruise down the length of the lake takes you through an almost complete succession of the Paleozoic formations found in the Canadian Rockies. Starting in slide-covered Triassic clastic strata of the Sulphur Mountain Formation surrounding the outlet of the lake, you pass successively by Carboniferous, Devonian and Ordovician formations until, at its southeastern end, the lake lies in strata of Cambrian age. The southwest-facing dip slope flanking the northeast shore (and connecting the ridges of Leah and Samson peaks and Mt. Paul) is formed from westerly inclined carbonate strata of Carboniferous and Devonian age. These rocks, together with older strata out of sight on the east side of the ridge, form the Chetamon Thrust Sheet, which has been folded into northwesterly trending anticlines and synclines. You may recall that we spoke of the Chetamon Thrust in Part One, in our discussion of how the front ranges consisted of a series of shingled sheets of strata, each separated by thrust faults. Along Highway 16, east of Jasper, this same fault crosses the highway close to the Snaring River Bridge.

Surrounding the southeastern end of Maligne Lake are such magnificent, ice-maned peaks as Monkhead, Mt. Brazeau, Mt. Mary Vaux, Llysyfran, Mt. Charlton and Mt. Unwin, all carved from carbonate and clastic rocks of Cambrian to Devonian age. Carbonate strata form the steep cliffs which, on most of these mountains, are commonly separated by gentler slopes of shaly layers. Several anticlines, synclines and minor thrust faults disrupt these strata, particularly on the peak adjacent to Mt. Mary Vaux.

Returning to the northwest end of the lake, we see the lush alpine meadows of the Bald Hills to the west. The contrast in topography between these gently rounded hills and the precipitous slopes across the lake is due to differences in rock type. The Bald Hills are underlain by Precambrian Miette Group shale, slate and gritstone, which erode much more easily than the tough, more resistant Paleozoic carbonates.

Looking northwesterly down the Maligne Valley, you can see the sawtooth Devonian carbonate peaks of the Queen Elizabeth Ranges, each range floored by a thrust fault and together forming an overlapping series of thrust sheets. The great tilted limestone slabs displayed in those thrust sheets are typical of the front ranges.

The Skyline Trail
Walking distance 44.1 km (27 miles)

The trail from Maligne Lake to Maligne Canyon via Shovel Pass and the Skyline Trail provides the most magnificent hike in the Canadian Rockies. Travellers generally take two or three days to complete the journey; however, the fitness nuts among you can run it in one. There are mosquito-infested campsites with **bear-poles** and latrines in the Snowbowl, near Curator Lake, at Tekarra Lake, and near the top of Signal Mountain. Campfires are not permitted, so backpackers must bring their own camp stoves. All garbage must be packed out by those who produced it. In 1994 Parks Canada began charging back-country user fees of five dollars per person per night. Backpackers must buy passes at the trail office in the Information Centre in Jasper. You may wish to make your reservations a week or two in advance, for the trail is so popular that the park authorities must limit the number of people in the region at any one time.

Upon leaving Maligne Lake, you cross the overgrown rockslide damming the lake and pass by a series of circular depressions. For many years these were thought to be **kettles** (hollows caused by calving of large blocks of ice off the front of a retreating glacier), but a recent study suggests that these depressions are still deepening and hence may be caused by the dissolution of limestone by water passing through the slide debris beneath the forest cover. If so, they are karst features.

Near Mona Lake you cross the surface trace of the Pyramid Thrust Fault, the fault separating the front ranges and main ranges. You can't see the fault, as it's covered by slide debris. A few kilometres farther you reach Evelyn Creek and begin a steady but pleasant climb to the crest of Little Shovel Pass. There are few outcrops on this part of the trail, but as you reach higher elevations a look back reveals marvelous views of the grey Devonian limestones of the Queen Elizabeth Ranges along the northeast side of Maligne Lake.

Cresting Little Shovel Pass, you begin your descent into the beautiful, mountain-rimmed alpine valley known as the Snowbowl. The surrounding mountains are carved from Precambrian slate and gritstone, the latter commonly forming the dark, lichen-encrusted ridges.

Jumbled **rock glaciers** spread out at the bases of these ridges (see upper photo, page 91). Rock glaciers are rather like ice glaciers: they move (albeit much more slowly than ice glaciers), as shown by concentric ridges on their surfaces. Ice between the boulders is responsible for their slow movement; however, those who have studied the rock glaciers of Jasper National Park think that most have lost their ice and are no longer moving.

Crossing the Snowbowl, you pass through the back-country campground at the entrance to the alpine meadows leading to Big Shovel Pass. Stretching out before you like the fairways of a huge golf course, the meadows in late July and early August are alive with colour: white and pink heather, purple larkspur, yellow arnica, red paintbrush, and, in small enclaves surrounded by scree on the higher slopes, the brilliant blue of tiny forget-me-nots. These and many other alpine flowers attract your attention, as does the occasional herd of mountain caribou that graze on the cream-coloured reindeer lichen in the vicinity of Big Shovel Pass.

The surrounding peaks, such as Antler Mountain and Curator Mountain, are composed of the same Precambrian Miette Group gritstone and slate viewed from the Snowbowl. However, the principal geological interest here lies in the structures into which these strata have been deformed. Anticlines, synclines and faults can readily be seen on the north-east faces of these mountains, particularly on Curator Mountain, where, from a lunch stop at Curator Lake, you can take your time to study the obvious effects of a thrust fault where it has cut across folded strata. Alternatively, you can enjoy the antics of the marmots that inhabit the clefts between the many boulders strewn about the shore. Look for a bed of apricot-coloured Precambrian dolomite near the lake.

After your rest, you face a gruelling hour's climb up to the Notch, the beginning of the truly "skyline" part of the Skyline Trail. Until late in the summer a steep bank of snow must be crossed just below the top of the Notch. This is all that remains of a **cornice** – snow that has blown over the top of the Notch during the previous winter and built up in a hardened wave-like shape.

From the Notch the trail parallels the surface trace of a thrust fault that occurs along the crest of the ridge as far as Mt. Tekarra. Several other

The alpine meadows leading to Big Shovel Pass and the Skyline Trail are rimmed with mountains carved from strata of the Precambrian Miette Group.

Photo by C.J. Yorath.

The view from the Notch on the Skyline Trail is spectacular indeed. Across the broad Athabasca Valley to the extreme left, the Cambrian quartzite prow of Mt. Edith Cavell thrusts boldly toward you above Precambrian Miette Group rocks supporting lower mountains in front. Behind, on the south side of the Astoria River valley, is Throne Mountain, and on the north side, in the far distance, the dark Cambrian cliffs of the Ramparts rise vertically above the Amethyst Lakes in the Tonquin Valley. Farther north to the right of Franchère Peak and Aquila Mountain, Portal Creek drains into the Athabasca. North of Portal Creek, ski runs scar the slopes of Marmot Mountain, which is carved from Precambrian Miette Group gritstones and slate. To the right are The Whistlers and the Jasper Tramway. At the extreme right lies the valley of the Miette River, and in the far distance, barely discernible as a white pyramid, is Mt. Robson, the highest peak in the Canadian Rockies.

Photo by C.J. Yorath

thrust faults dislocate Miette strata along the slope beneath the ridge. Across the Athabasca Valley are Mt. Edith Cavell, Throne Mountain, Marmot Mountain and The Whistlers. Farther beyond you can see the Ramparts forming the western wall of the Tonquin Valley (see page 150 and lower photo on page 98). Tall peaks of the Columbia Icefield form the southeastern horizon (see pages 126–32 and photos on pages 94 and 95), while to the northwest, on a clear day, you can see more than 80 km to the glacier-whitened summit pyramid of Mt. Robson, highest peak in the Canadian Rockies (see pages 49–51 and photo on page 85).

Watch for patches of **patterned ground** along this bare, stony ridge. The soil is permanently frozen at this elevation, a condition known as **permafrost**, even though the top 20 cm or so melts each summer. Daily freezing and thawing imparts a churning action to this layer that tends to segregate the larger rock fragments from the smaller ones, creating a pat-

In permafrost areas the daily freezing and thawing of the upper few centimetres of rocky soil result in the formation of stone polygons where the surface is horizontal (top). On sloping surfaces the polygons stretch out downslope to form stone stripes (bottom).

Photos by Ben Gadd

From the lovely alpine meadows on the slopes of Signal Mountain, the broad sweep of the Athabasca Valley can be seen extending eastward toward the mountain front. To the far left is Pyramid Mountain, formed of quartzite of the Cambrian Gog Group. Next to it are the carbonate cliffs of the Palisade.

Photo by C.J. Yorath

tern of geometrical shapes on the ground. If the surface is flat, polygons are created, with fine fragments in the centres of the polygons and coarse fragments at their edges. If the surface is sloping, the polygons stretch out downhill and become **stone stripes.** You can see both types of patterned ground along the trail.

Just above the Centre Lakes, opposite Centre Mountain, the trail descends through many switchbacks into the valley of Tekarra Lake. Watch for **solifluction lobes** as you reach the lush, green tundra of the lower slopes. A solifluction lobe is another permafrost feature, this one dependent on fine-grained soil and a thick covering of vegetation. As the surface layer thaws, the water-saturated soil becomes unstable. Were it not for the mat of tundra plants holding it together with their shallow roots, the soil would flow downhill as mud. Because the zone of melting extends somewhat deeper than the roots, the mud below the root zone moves, carrying the vegetation mat with it. The mat rolls under the front of the lobe as it advances down the hill; the bend in the mat at the lobe's front is stiff enough to give the solifluction lobe a steep face up to 2 m in height. Solifluction lobes form most readily in alpine meadows on cold northeastern slopes just above the timberline, and this is the situation along the section of the Skyline Trail between Tekarra Lake and the Signal Mountain Fire Road.

If you camp near Tekarra Lake, the many fearless marmots of the region will beg for some of your food. Hope for a clear dawn, which will give you an outstanding sunlit view of the Cambrian Gog quartzite forming the main buttress of Mt. Tekarra; the lower, less resistant slopes are Precambrian Miette Group strata (see lower photo on page 91). After crossing Tekarra Creek, the trail leads into beautiful alpine country on the north sides of Mt. Tekarra and Signal Mountain. This is one of the joys of the trip, for the views of the geology surrounding the Athabasca Valley are breath-taking: the reddish Cambrian quartzite bulk of Pyramid Mountain riding the thrust fault named for it; the Paleozoic carbonates of the Palisade, with its great cliff of Palliser limestone; the **braided,** glacial Athabasca River glinting in the sun as it swings north between the Palisade and the limestone slabs of the Colin Range; and far down the valley the craggy visage of Gargoyle Mountain, which wears a smile of snow on the syncline in its south face.

All too soon, it's over. The last 8 km are a tedious, foot-numbing slog down the abandoned Signal Mountain Fire Road to the Maligne Road near Maligne Canyon.

THE ICEFIELDS PARKWAY (HIGHWAY 93)
Jasper to Sunwapta Pass
Road distances given are from the intersection of Highway 16 and Highway 93.

Travelling south from Jasper along the Icefields Parkway on a clear, sunny day in any season is one of North America's great treats. In late autumn, brightly coloured aspen leaves fly in the afternoon breeze, rising in eddies against the backdrop of the grey and umber-coloured mountains. Elk descend from alpine meadows. Chipmunks fill their underground larders along the banks of the Sunwapta River. Soon, winter's whitest of whites beneath bluest of blues will lend a crisp starkness to the shapes of the peaks forming the Winston Churchill Range to the west and Endless Chain Ridge to the east.

(22) Athabasca Falls
Turnoff 29.9 km, 18.6 miles

From Jasper to Athabasca Falls the highway passes through outcrops of slate, gritstone and pebble conglomerates of the Precambrian Miette Group. The Maligne Range, east of the highway, is also mostly Miette Group, but the high peaks west of the highway are cut from quartzite of the Cambrian Gog Group. At Athabasca Falls, between the knife-edge ridge of Mt. Edith Cavell to the west and the dish-shaped synclinal form of Mt. Kerkeslin to the east, the Athabasca River boils through a gorge cut into these hard lower Cambrian strata. (See photos on pages 92, 94 and 97.)

(23) Goats and Glaciers Viewpoint
36.2 km, 22.5 miles

This stop is aptly named, for here it is often possible to see both mountain goats and glaciers. The view across the Athabasca Valley is a classic glacial landscape, dominated by the horn peak of Mt. Fryatt (3361 m). The cliffs there are the backwalls of cirques, several of which hold small glaciers. A larger glacier lies on a rock shelf; 150 years ago, at the height of the Little Ice Age, it reached the lip of its hanging valley before melting back during the recent glacial retreat. Most of the ridges are arêtes, knife-edge crests whittled thin by glaciers cutting adjacent cirques.

Lower down, note the broad terrace in the treed slope above the river. The viewpoint is built on what remains of that same terrace on this side of the river. Looking down to the river, you can see that the terrace is made of fine, silty glacial **rock flour.** This was carried by the Athabasca River and dumped when glacial ice or related deposits blocked the valley,

Glacial rock flour provides necessary mineral nutrients for mountain goats at Goats and Glaciers Viewpoint along the Icefields Parkway (Highway 93).

Photos by Ben Gadd

Mt. Fryatt and its neighbouring peaks are carved from horizontal strata of the Cambrian Gog Group.

Photo by C.J. Yorath

creating a dam that backed up the river. The resulting lake's quiet water allowed the rock flour to settle out of the water, gradually filling the lake basin until the river could flow freely across it, leaving sand and gravel as the uppermost layer of the deposit. Loss of the dam, whether by glacial melt or river erosion, allowed the river to cut down through the rock-flour deposit, leaving the terrace on each side.

Mountain goats come down from the cliffs of Mt. Kerkeslin (see photo on page 92), the large, colourful peak across the highway, to lick this rock flour. Grazing animals crave sulphur and sodium, essential to the growing of new coats each year. The females need phosphorus to form the bones of their young, and they must consume a great deal of calcium to produce milk for nursing. The rock flour here, being glacially derived from rocks rich in these and other minerals, is easily exploited by these grazing animals.

Mt. Fryatt and the neighbouring peaks to the southwest invite you to don boots and a backpack and hike into the high, hanging alpine valley of Fryatt Creek. The route is surrounded by a crowd of Cambrian peaks mantled by glaciers. The trail starts at Athabasca Falls.

Close to the point the **Sunwapta River** joins the Athabasca, a steep, narrow gorge has been cut through the Cambrian Cathedral Formation. At Sunwapta Falls the Sunwapta River cascades into a plunge pool, undercutting the lip of the falls and causing progressive upstream erosion of the limestone.

Photo by C.J. Yorath

(24) Sunwapta Falls
Turnoff at 53.0 km, 32.9 miles

At Sunwapta Falls the Sunwapta River has cut a gorge through Cambrian limestone of the Cathedral Formation. This is where the valley of the Sunwapta River joins the valley of the Athabasca River. During the ice ages the Sunwapta Valley glacier was smaller than the Athabasca Valley glacier, and thus the Sunwapta Valley was not cut as deeply as the latter. Just as important was the sideward-cutting power of the Athabasca Valley glacier, which, in widening its valley, trimmed away the downstream end of the Sunwapta Valley. The result after the ice melted was a drop of 50 m at the mouth of the Sunwapta River, increasing the river's speed and erosive power. The channel quickly developed a series of steps, one of which grew at the expense of the others and became the main waterfall. The plunge pool at the base of the waterfall undercuts the lip of the falls; the lip breaks off repeatedly, causing the falls to erode their way upstream, leaving a gorge in their wake.

Also of interest here is the right-angle bend in the gorge, where the river has followed the jointing in the rock.

(25) Endless Chain Ridge, Bubbling Springs and the Jonas Slide
59.0 km, 36.7 miles

Not far past Sunwapta Falls you begin to pass along the base of Endless Chain Ridge, so named by early wilderness traveller Mary Schäffer, whose horse party passed under the ridge's monotonous chain of summits for three days in 1907. Her account of that trip does not mention Bubbling Springs, an attractive little pool at the base of the ridge. The rate of flow of the crystal-clear groundwater moving up through the bottom of the pool is steady and fast enough to wash all the tiny clay- and silt-sized particles from the sediments in the pool, but slow enough to leave those of sand size. Larger fragments have sunk into the sand.

Air bubbles rise from many spots on the bottom where groundwater flows into the pool. The presence of air carried in the flow of the water suggests that the groundwater hasn't travelled very far underground. It probably runs down from the quartzite slabs above, perhaps entering a small cave or fracture in the thin layer of limestone overlying the quartzite near the valley floor, and exiting under a small amount of pressure soon thereafter. The water squirts up briskly, forming small fountains of sand. Watching the pool for a while shows that different fountains are active at different times.

Endless Chain Ridge is a dip slope of Lower Cambrian Gog Group quartzite. As we saw at Medicine Lake and Maligne Lake, such slopes, which follow the inclination of the strata, are prime landslide areas. Sheets of jointed rock, oversteepened by glacial undercutting at the floor of the valley and loosened by frost action or water flowing between the beds, break loose to careen down into the adjacent valleys. Jonas Slide *(73.2 km, 45.5 miles)* was just such a landslide (see photo on page 93). The highway had to be bulldozed through the debris. An estimated 15,600,000 m^3 of pink Gog quartzite broke away in two slides, the **crowns** (crescent-shaped release points) of which are clearly visible high up on the slope of the nearby ridge. Two sets of joints are present here, both perpendicular to the stratification, thus accounting for the blocky nature of the slide debris.

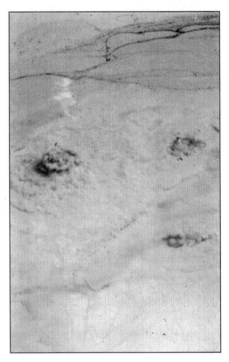

Fine sand on the bottom of a crystal-clear pool at Bubbling Springs forms tiny plumes as water jets up from underground. Escaping air, carried in the water, produces the bubbles.

Photo by Ben Gadd

If you clamber over these blocks, you will see excellent examples of **cross-bedding** in the quartzite boulders. The layers intersect at different angles, indicating that the sand was swept along by shifting currents, depositing sand in one spot and eroding it in another, just as occurs in the delta of a modern river. The pink colour comes from weathering of tiny iron pyrite crystals in the rock.

The dip slope of Endless Chain Ridge is formed from westerly inclined strata of the Cambrian Gog Group.

Photo by C.J. Yorath

(26) Beauty Creek Flats
85.8 km, 53.3 miles

A rockslide blocking a river has an important effect on the appearance of the valley upstream, as we see here at Beauty Creek Flats. The Jonas Slide was large enough to dam the river temporarily, producing a lake. Because the Sunwapta River carries large quantities of gravel from its glacial source at the Athabasca Glacier, it must have filled in the lake quickly, perhaps in only a couple of thousand years. This would have reduced the gradient of stream course and the stream's **competence** (the ability of the stream to carry its sediment load), with the result that the river has deposited most of its material and in so doing has created the flats. The growing alluvial fan of Diadem Creek, at the north end of the flats, is now forming another blockage in the valley.

From its emergence at the toe of the Athabasca Glacier the Sunwapta River displays the wide variation in flow typical of glacial rivers. The summer maximum flow is about thirty times that in winter, and, additionally, there is a wide daily variation from low flow rates late at night (when glacial melt is reduced) to high rates in the late afternoon (when melt is greatest). A combination of these factors has given the river its many shallow channels. The annual summer pulse of water acts as a flood, modifying the previous fall's channel courses. When the river is running at its daily maximum, it has the power to pick up sediment, in this way cutting new channels. When it slows to its daily minimum, it loses competence and tends to deposit sediment, thus plugging its recently cut channels. By these means the channel pattern changes day by day, as well as year by year, becoming braided.

The Sunwapta River braids its way down a wide, gravel-floored glacial trough abandoned only 12,000 years ago by the ice that carved it. Not far upstream lies the Columbia Icefield, largest glacier in the Rocky Mountains, whose meltwaters carry thousands of tonnes of alluvium down the steep gorge of Sunwapta Canyon and out onto the gently sloping Beauty Creek Flats, where the sand and gravel are dumped.

Photo by Ben Gadd

㉗ Tangle Ridge
87.9 km, 54.6 miles

Farther on, near the base of the long climb up to the Athabasca Glacier, you pass by the northwestern end of Tangle Ridge, where the surface trace of the Mt. Coleman normal fault forms the prominent gully on the northwestern slope of the ridge. The peak is supported by resistant, cliff-forming Ordovician carbonates of the Skoki and Outram formations, which, as far down as the top of the gully, are underlain by shaly rocks of the Survey Peak Formation. From there to the base of the ridge, a comparison of the geology on either side of the gully reveals one of the important criteria for identifying a normal fault: younger rocks on one side have been dropped down alongside older rocks on the other side. In this case, the Mistaya, Bison Creek and Lyell formations of Late Cambrian age, all on the southeast (right) side of the fault, have been down-dropped against the older Middle Cambrian Arctomys, Eldon and Pika formations on the northwest (left) side. Because of the presence of a peculiar pattern of folds in the Lyell Formation close to the fault, Ray Price of Queen's University believes that the Mt. Coleman Fault began life as a thrust fault and later changed its mind, to retire as a normal fault.

At the northwestern end of Tangle Ridge the Mt. Coleman normal fault follows the steep gully seen on the mountain's lower cliff. The top of the mountain consists of Ordovician carbonates and shaly rocks. In the lower cliffs, rocks of Late Cambrian age on the right of the gully have been dropped downward against older strata of Middle Cambrian age.

Photo by Ben Gadd

Viewpoints on the Tangle Hill

There are three fine viewpoints along the highway where it climbs steeply to Icefield Centre, the visitor-services operation near the Athabasca Glacier.

The Stutfield Viewpoint (**stop** (28), *92.4 km, 57.4 miles*) provides a magnificent view westward to enormous mountain walls of dark Middle Cambrian limestone and dolomite capped by white glacial ice. The two high points are Mt. Kitchener on the left and Mt. Stutfield on the right. Between them is an **icefall,** the place where the Stutfield Glacier, a part of the Columbia Icefield, flows over a cliff.

Everything here is on a giant scale. If you're willing to wait for a while, you may see a chunk of the glacier perhaps as large as an apartment building break off from the ice-front and tumble down the slope. Starting from 7 km away, the thunderous sound arrives well after you see the icefall; if a very large chunk drops off, clouds of ice chips billow out from the base of the cliff (see lower photo on page 95). Enough ice chips and snow accumulate at the foot of the icefall to reconstitute the glacier, which flows for a couple of kilometres toward the Sunwapta River.

At the Tangle Falls Viewpoint (**stop** (29), *94.1 km, 58.5 miles*), a lovely waterfall splashes over low, buff-coloured cliffs of Upper Cambrian Mistaya Formation limestone. The Mistaya, like most of the other Upper Cambrian units in the Rockies, contains **stromatolites** – plate-size, domelike masses in the rock layering created by colonies of algae. An exposed bedding surface showing the tops of stromatolites looks like a puffy quilt.

At the Mount Kitchener Viewpoint (**stop** (30), *94.9 km, 59.0 miles*) you can see the effects of another landslide that blocked the Sunwapta River. The Kitchener Slide, with an estimated volume of 39,000,000 m^3, brought a good-sized mass of the mountain down into the valley, filling it to a depth of 150 m with fragments of the Middle Cambrian Eldon Formation. The Sunwapta River has cut a deep gash through the slide, incising it down to and into the bedrock of the valley floor beneath. If you look closely, you can see water dribbling out at the base of the rubbly slide, then streaming down the smooth wall of the bedrock gorge.

As in the case of the Jonas Slide downstream, another area of gravel flats has developed on the upstream side of the Kitchener Slide, where the highway descends on its approach to the Athabasca Glacier. The difference between these and the flats at Beauty Creek is that the Kitchener Flats are marked by low terraces on each side of the river. This indicates that the river, which once deposited that gravel, is now cutting into it. Geologists trace the change in erosive power to the creation of Sunwapta Lake at the toe of the Athabasca Glacier. Formed in the late 1930s as the glacier's

A tangle of waterfalls cascade over buff-coloured carbonate strata of the Upper Cambrian Mistaya Formation at Tangle Falls.
Photo by Ben Gadd

Across the highway from the Mount Kitchener Viewpoint a spectacular array of calcite-filled tension cracks occurs within shaly limestone of the Upper Cambrian Bison Creek Formation.

Photo by C.J. Yorath

retreat exposed a depression in the valley floor, the lake now collects sand and gravel that used to flow down the river. The river, now lacking this sediment load, has more erosive power than it used to have, and will thus cut its way deeper into the flats until Sunwapta Lake fills with debris from Athabasca Glacier.

Across the highway from the Mount Kitchener Viewpoint, the effect of tension (stretching) on rock is clearly demonstrated. There one can see two sets of parallel white streaks in strata of the Upper Cambrian Bison Creek Formation. The streaks are formed from the mineral calcite ($CaCO_3$), which filled small tears in the rock as it was pulled apart.

(31) Athabasca Glacier and the Columbia Icefield
101.3 km, 62.9 miles

It's important to realize that what you see here is not the Columbia Icefield itself. The icefield is out of sight at the head of the Athabasca Glacier.

The Athabasca Glacier (see upper photo on page 94) is one of some thirty separate glaciers draining the central ice mass of the Columbia Icefield, together forming a glacier complex covering an area of more than

250 km² (see the geological map). Climbers reaching the summit of Mt. Athabasca, the big, ice-covered peak to the left of the Athabasca Glacier, can see most of the Columbia Icefield. It looks like a huge white blanket held up by the peaks along its edges: Mt. Athabasca, Mt. Andromeda, Castleguard Mountain, Mt. Columbia, North Twin, Stutfield Peak, Mt. Kitchener and the Snow Dome. The average elevation of the icefield is 3000 m above sea level, and two of its peaks, North Twin and Mt. Columbia, are over 3700 m in elevation.

The Columbia Icefield is the hydrographic apex (a point away from which rivers drain) of North America, the only point from which water flows to all three oceans. Earlier we noted that the southern boundary of Jasper National Park is at a continental divide that separates rivers draining to the Arctic Ocean from rivers draining to the Atlantic Ocean at Hudson Bay. This east/west divide joins the north/south divide at the summit of the Snow Dome on the Columbia Icefield, making the Snow Dome the hydro-

The plateau-like nature of the Columbia Icefield is obvious in this picture. The photo shows only about half the 300 km² ice mass, which is the largest in the Rocky Mountains. This view is to the northeast, with Mt. Columbia (highest peak in Alberta, 3747 m) at left and the gently rounded Snow Dome (3520 m) on the skyline at centre. The Snow Dome's summit is the hydrographic apex of North America, from which water flows to three oceans: Atlantic, Pacific and Arctic.

Parks Canada photo

graphic apex of the continent. Ice flows away from the summit of the Snow Dome toward the Athabasca Glacier (Arctic drainage via the Sunwapta, Athabasca, Slave and Mackenzie rivers), the Saskatchewan Glacier (Atlantic drainage via the North Saskatchewan, Saskatchewan and Nelson rivers), and the unnamed glaciers flowing off the southwest side of the icefield (Pacific drainage via Bryce Creek and the Bush and Columbia rivers). The summit of the Snow Dome is thus the only point in North America from which a mountaineer can pollute three oceans in one act.

The Athabasca Glacier flows over Cambrian rocks above the Simpson Pass Thrust. Since the mid-1840s, when the Little Ice Age reached its maximum after a 600-year buildup, this best-known glacier in Canada has been retreating. Its former extent can be appreciated if we imagine that ice fills

One of several recessional moraines of the Athabasca Glacier is cut by the roadway leading to the modern glacier. In the middle ground of the photo the glacier's lateral moraine lies at the foot of Mt. Athabasca.

Photo by Ben Gadd

These parallel scratches – glacial striations – on bedrock indicate the direction in which the glacier moved.

Photo by Ben Gadd

The striped appearance of "tiger rocks" is due to alternating layers of dark grey limestone and orange-weathering dolomite.

Photo by Ben Gadd

the entire valley between the two lateral moraines and covers the highway where it passes by the buildings of Icefield Centre.

The melt-back area, properly termed the **forefield,** has been studied by Brian Luckman of the University of Western Ontario. Its many features include an **end moraine,** the low gravel ridge marking the point of farthest advance; **recessional moraines,** each ridge showing where the glacial front stalled for a few years before melting back farther; **ground moraine,** a blanket of boulders and stones that once stuck to the glacier's underside and did the work of scraping away at the valley floor; **striations,** or scratches, on bare bedrock surfaces that point in the direction the glacier flowed; and "tiger rocks," boldly striped orange-and-grey boulders of interbedded Cambrian limestone and dolomite.

Brian has taken cores of trees growing inside and outside the forefield and counted their growth rings to determine their age. In so doing he discovered a grove of gnarled Engelmann spruce old enough to have lived during the Little Ice Age; he has used this and other information to work out the melt-back sequence of the Athabasca Glacier.

For those of you who want to know more about the modern glacier we recommend an excellent book entitled *Exploring the Columbia Icefield* by Dick Kucera. The book can be purchased at the literature sales desk in the visitors' centre.

The following are some essential facts about the Athabasca Glacier. It is 6 km long and 1 km wide. The thickness of the ice in the main mass, as measured by geophysical techniques, reaches 365 m. It starts at an elevation of 2700 m and ends at an elevation of 1950 m. Flowing down the valley at 15 to 25 m per year, depending on what part of the glacier you measure, the ice takes 150 to 200 years to make the trip. The rate of melt at the toe of the glacier is presently faster than the rate of forward advance, thus the toe is retreating at about 20 m per year. At the same time, the surface is melting downward at a rate of approximately 3 m per year. Dick Kucera believes that the glacier will soon melt back behind a bedrock ridge that crosses the valley floor, uncovering a basin and thus creating a lake that icebergs will drop into from the glacier's snout.

The mountains surrounding Athabasca Glacier are carved from Cambrian carbonates and limy shale. On the northeast slope of Mt. Athabasca, the surface trace of the Simpson Pass Thrust crosses the mountain between the ice-covered flank and the bare ridge in front of it. Look for outcrops of red and green shales beside the highway where it turns north just past Icefield Centre; these belong to the Cambrian Arctomys Formation, which contains casts of salt crystals (salt hoppers) indicating that the ancient continental shelf went through a period of drying about 530 million years ago. The rocky peak behind Icefield Centre is Mt. Wilcox. Its lower cliffs display the Cambrian Lyell Formation.

Northeast of Icefield Centre is Nigel Peak, on the boundary between

The surface trace of the Simpson Pass Thrust passes through the shallow notch on the left slope of Mt. Athabasca between the ice-covered northeast face and the bare ridge in front of it.

Photo by Ben Gadd

Cube-shaped casts of salt crystals, called salt hoppers, in the Arctomys Formation near Icefield Centre offer evidence of a period when the Cambrian seafloor of western Canada was drying (about 530 million years ago).

Photo by Ben Gadd

Banff and Jasper national parks. The mountain's peak is carved from Carboniferous limestones and dolomites of the Rundle Group, which are successively underlain by shales of the Banff and Exshaw formations (see below) that form the lower slopes and finally by massive limestone of the Devonian Palliser Formation. When viewed from the viewpoint near Nigel Creek (**stop** (32), *113.8 km, 70.7 miles*) on the big hill south of Sunwapta Pass *(105.9 km, 65.8 miles)*, the strata of Nigel Peak form a syncline extending from near Tangle Ridge, through Nigel Peak and on to Mt. Wilson on the North Saskatchewan River. The Mt. Coleman normal fault, also seen at Tangle Ridge (**stop** (27)), extends southward to Nigel Peak, where it separates the above-mentioned Carboniferous strata on the west from Ordovician and Cambrian rocks on the east. Seen from the Big Hill, the surface trace of the fault crosses the valley between Nigel Peak and the mountain to the southeast.

The Exshaw Formation is worthy of comment. Throughout the Rockies it consists of a thin succession of black, organic-rich shale lying between the Devonian Palliser Formation and the Carboniferous Banff Formation. The same formation, known by different names, occurs throughout much of North America, where it rarely is more than 6 m thick. It is probable that the sediments of the Exshaw accumulated in huge, shallow lagoons where the organic carbon remains of organisms could be preserved in poorly oxygenated waters. That such conditions existed over such a vast area of the continent is indeed puzzling. An important aspect of the Exshaw Formation is that its organic component probably supplied much of the hydrocarbons for Carboniferous oil and gas reservoirs throughout the foothills of Alberta.

If you look north from a viewpoint on the big hill south of Sunwapta Pass, you can see that the Mt. Coleman normal fault passes through the valley separating synclinal Devonian and Carboniferous strata forming Nigel Peak on the left and Cambrian and Ordovician rocks supporting the mountain on the right.

Photo by C.J. Yorath

Other Localities of Geological Interest

(33) Jasper townsite area

About 12,000 years ago, deposits of sand, gravel and boulders accumulated across the floor of the Athabasca Valley during the melting and retreat of the Cordilleran Ice Sheet. Since then, the river has carved down into these deposits, producing terraces. The town of Jasper is situated on one of these terraces, and Jasper Park Lodge sits on a lower terrace on the other side of the valley.

The bedrock beneath these deposits consists of folded blue, green and purple slate and pink limestone of the Precambrian Old Fort Point Formation, the oldest formation of the Miette Group in this area, and the somewhat younger gritstone and slate layers of the middle Miette Group, seen along the Yellowhead Highway west of town.

On their geological map of the Jasper townsite area, Eric Mountjoy and Ray Price have shown four faults extending southeasterly from the Victoria Cross Ranges and cutting through the bedrock beneath the town.

An early-morning view from the top of Old Fort Point of the town of Jasper. The town was built upon a river terrace of sand and gravel that spread over the floor of the Athabasca Valley during the melting and retreat of the Cordilleran Ice Sheet about 12,000 years ago. Cambrian quartzite strata of the Gog Group form the peaks of the Victoria Cross Ranges in the background.

Photo by Ben Gadd

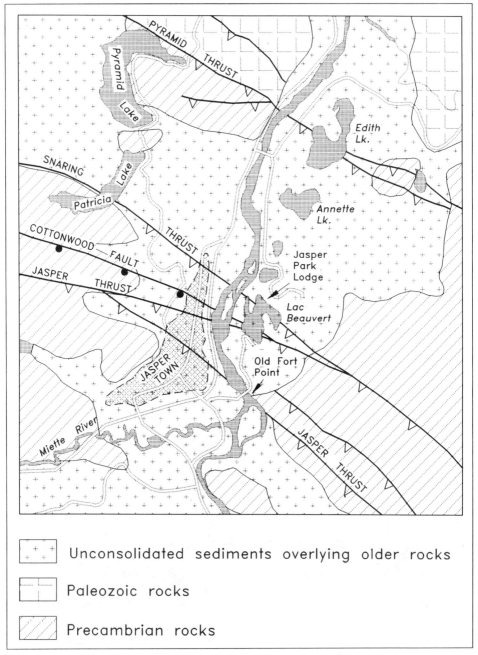

This geological map of the Jasper townsite area shows where four faults pass through the bedrock beneath the town.

Based on a map published by E.W. Mountjoy and R.A. Price

The most northerly is a thrust fault, part of the Snaring Thrust, which crosses beneath the north end of town and continues under Lac Beauvert and the Jasper Park Lodge golf course. To the south a normal fault, the Cottonwood Fault, passes beneath the Lobstick Lodge on Geikie Street and continues through the south end of Lac Beauvert. Farther south are two thrust faults. The more northerly, a splay off the one to the south, passes beneath Ben Gadd's house on Geikie Street and continues southeasterly to join with the Snaring Thrust east of Lac Beauvert. The southerly fault of the two is the Jasper Thrust, which passes under the high school on Pyramid Avenue and continues along the south face of Old Fort Point, a prominent hill that forms the eastern bank of the Athabasca River.

The presence of these faults is no cause for alarm. There is no evidence that they have been active since mountain-building ended some 60 million years ago. Moreover, no earthquakes greater than magnitude 2 have yet been recorded anywhere in the park since the early 1960s, when the Geological Survey of Canada began to monitor earthquakes in the region. Prior to that, only large-magnitude events, meaning those actually felt, could have been recorded. No such events are known to have occurred.

A stroll along Connaught Drive, Jasper's main street, provides open views to the east, across the Athabasca Valley to Devonian and younger strata in the front ranges. These sediments accumulated upon the shallow shelf of western North America between 375 and 200 million years ago. Unmistakable among the peaks is the sombre, upward-gazing, snow-covered face of "The Old Man," also known as Roche Bonhomme (upper photo on page 95; see page 59 for a discussion of the view of its northeast side). The Old Man's eyebrow and nose are carved from Triassic siltstone; his large jaw is made of Permian chert. The Old Man rests upon a pillow of Carboniferous carbonate strata of the Rundle Group. Successively lower strata belong to the Carboniferous Banff and Upper Devonian Palliser formations. These three stratigraphic units – Palliser carbonates, Banff shales and Rundle carbonates – form a three-part succession of formations that is repeated, ridge after ridge, as the characteristic, unique signature of the front ranges of the Canadian Rockies. Where they occur in completely exposed successions, such as in the cliffs of Mt. Rundle and Cascade Mountain in Banff National Park, they are easily recognized. The Palliser and Rundle carbonates form steep, pale-grey cliffs, whereas the middle shaly rocks of the Banff Formation are more easily weathered and eroded, appearing as rubbly dark slopes between the lower and upper cliffs.

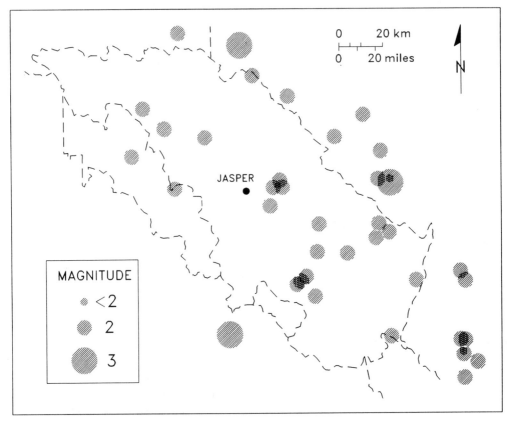

Epicentres of earthquakes recorded in the vicinity of Jasper National Park between 1960 and 1991. The magnitudes of these earthquakes are shown by the sizes of the circles. Within Jasper National Park all of the recorded events are of magnitude 2, too low to have been felt unless one was standing very near any of the epicentres. The magnitude 3 events could have been felt by people within 30 km of the epicentres, but all of these occurred outside the park boundaries. The causes of these low-magnitude events are probably related to the presence of residual stresses still remaining in the crust as a consequence of mountain-building, which ended about 60 million years ago.

Courtesy of R.B. Horner, Geological Survey of Canada

(34) Marmot Mountain

From its intersection with Highway 93, take the Cavell Road until you come to the road leading to the Marmot Basin ski area, which turns off to the right. Follow the Marmot Road to one of several parking lots at the ski development (see upper photo on page 98).

The big chairlift has just dropped you off and you are standing on the eastern slope of Marmot Mountain. Skis parallel. Ankles together. Knees slightly flexed. Poles held lightly in front of you, their tips barely touching the snow. It is early spring. The air is crisp and clear. An electric blue sky surrounds a brilliant sun, its reflection gleaming from the powder. You savour the faint, intoxicatingly sweet scent of sap rising in the lodgepole pines. As you look far below, the broad sweep of the Athabasca Valley carries your gaze from Pyramid Mountain to the Columbia Icefield.

Beneath you, beneath the snow, is Precambrian dark grey and black silty shale of the Miette Group. As you glide downward from the top of the lift, you pass over a thrust fault, the Simpson Pass Thrust, one of the major structures of the Rockies. Crossing over the fault, you descend first onto coarse-grained clastics of the Cambrian Gog Group and then suddenly back onto Precambrian strata farther down the slope. With the powder flying away in your wake, you slice your way down the mountain across another fault, this time a normal fault that broke the crust in a relaxing sigh of relief after all of the pushing and shoving ceased about 60 million years ago. On the other side of the fault is a layer of gritstone, made from sediments carried by rivers from the ancient Precambrian Shield to the east and laid down across the torn and jagged edge of the continent formed when Australia was rifted away from North America about 750 million years ago. Back into the liftline and up you go again.

(35) Pyramid Mountain

From the junction of Pyramid Avenue and Pyramid Lake Road in Jasper, take the Pyramid Road up the hill and follow it *4.6 km (2.9 miles)* west, first to Patricia Lake, then to Pyramid Lake. Across the lake to the northwest is Pyramid Mountain (photo on page 96), elevation 2763 m, its peak regrettably reduced by 10 m to make room for a microwave relay station.

This peak, lying at the southeastern end of the Victoria Cross Ranges, is composed of northeasterly inclined, rusty red and brown quartzite, sandstone and conglomerate strata of the Cambrian Gog Group above the Pyramid Thrust. Along the fault, Cambrian and Precambrian rocks have been thrust eastward and over younger Paleozoic strata. The break in slope near the top of the mountain marks the trace of the Snaring Thrust, which, together with the Pyramid Thrust, isolates the peak from the remainder of the mountain, forming a klippe.

(36) Old Fort Point

A walk across the glacially eroded bedrock summit of Old Fort Point is a profound delight, winter or summer. From its crest, 120 m above the Athabasca Valley floor, splendid views of the environs of Jasper are available. It is a tranquil place, giving one a sense of belonging to the planet, to the universe.

From the junction of Highway 93A and Highway 16, proceed south about 100 m, then turn left (east) and continue 800 m to the parking area beyond the bridge over the Athabasca River. There are two ways to walk to the summit. The quicker (and steeper) route begins at the staircase at the base of cliff and continues on the trail above, which reaches the top in less than a kilometre. For the longer, gentler route, follow trail number 1A east into the forest. The trail parallels the cliff base for 2 km, then swings south and finally back to the east, gaining elevation through aspen groves, pine woods and wildflower meadows, reaching the summit after 5.5 km. From there it's a ten-minute walk down the quick-ascent route to the parking lot.

The hill is carved from grey, green and purple slate, siltstone, and pale grey-green limestone of the Old Fort Point Formation of the Miette Group, the oldest rocks in Jasper National Park. Careful inspection of the cliff by the parking lot reveals several interesting features, as shown by the photographs on pages 139–41. The strata are steeply inclined toward the southwest, where they form the flank of an anticline-syncline pair of folds. Notice how the strata are **cleaved** into thin slaty sheets and, moreover, that the southwestward inclination, or dip of the cleavage, is less than that of the stratigraphic layering. If you look along the dip slope forming the bank of the river, you will see that the rocky bank is banded in nearly horizontal layers of different-coloured tones. The banding is due to the intersections of the planes of cleavage with the planes of stratification, the latter forming the rocky bank.

Farther along the river's shore there is an abrupt change in rock type, where vertically dipping dark slates rest beside those we have been discussing. The contact between these two types of rock, each of which is part of the Old Fort Point Formation, is the Jasper Thrust. On Eric Mountjoy and Ray Price's geological map, the Jasper Thrust extends southeasterly from near Saturday Night Lake, through the town, to beyond this location.

Returning to the cliff by the parking lot, you can observe a 2 m wide, steeply inclined layer that has been broken up into many angular fragments of limestone, each fragment surrounded by brown siltstone. This breccia may have been caused by the sudden movement of still-soft sediment from shallow into deeper water, during which the layers were broken into individual fragments and enclosed within fine silt. Such movement may have been caused by earthquakes, which at that time, about 750 million years ago, would have been common as western North America rifted away from Australia.

To the left of this bed is another one much like it, but here the fragments of limestone have been dissolved out of their enclosing limy siltstone, such that rectilinear holes are all that remain (upper photo, page 141). Notice, too, that the limestone beds are cut by white veins of calcite.

Unlike the layers of slate and siltstone, the limestone strata are not cleaved. Indeed, you can readily see in the lower photo on page 141 that the planes of cleavage in the clastic strata end abruptly at the edge of the limestone bed. The limestone, being stronger, resisted being deformed in this manner, but the softer clastics yielded to the stress and developed cleavage.

Thin slaty sheets of the Old Fort Point Formation of the Precambrian Miette Group are inclined, or dip, toward the southwest at an angle of about 45°, whereas the stratification is nearly vertical, as shown by the smooth bedding surfaces at the lower right and upper left.

Photo by C.J.Yorath

Along the steep rocky bank of the Athabasca River the horizontally banded appearance of the rocks is due to the intersection of planes of cleavage with planes of stratification. At the extreme right the trace of the Jasper Thrust can be seen where vertically inclined dark slates forming the low bank rest beside lighter-coloured slates forming the bulk of the hill.

Photo by C.J. Yorath

Fragments of limestone (pale grey) surrounded by siltstone (brown and orange) form a layer of breccia. This feature may have been caused by the shattering of horizontal carbonate layers during an earthquake about 750 million years ago.

Photo by C.J. Yorath

Close by, a similar layer of breccia is riddled with holes where rectangular limestone fragments have been dissolved away.
Photo by C.J. Yorath

Slates of the Old Fort Point Formation, inclined at about 45°, are abruptly truncated by the more steeply inclined breccia layer, which, because it is stronger than the surrounding clastics, resisted the deforming forces that produced the cleavage in the softer slates.

Photo by C.J. Yorath

(37) The Whistlers

Rides on the Jasper Tramway are always fun. Some are even memorable. On one occasion the young attendant was explaining to her passengers that the mountain below was older than the mountains across the valley and that they were older than the mountains in the far distance. She had confused the age of the rock with the age of the mountain-building. When one of us volunteered to correct her, she said, "Every time I get a geologist in this crate I get a different story." We have since remained silent on these occasions.

From the upper terminal the trail to the summit of The Whistlers is a steep but short walk, and well worthwhile. The mountain is formed of Precambrian Miette Group strata that have been folded into anticlines and synclines. A cross-section of a small anticline can be seen on the face of a cliff to the west (right) as you start up the trail.

Outcrops and scattered boulders of gritstone, slate and rare carbonate are hiding places for marmots, their high-pitched whistling calls giving the name to the mountain. The large quartzite boulders found where the

From atop The Whistlers one can look far southward over the Athabasca Valley toward Mt. Kerkeslin and the Fryatt Group in the far distance. The Cambrian clastics of Mt. Edith Cavell form a gently dipping panel of strata on the right. In front of Cavell is a wide upland of beautiful alpine meadows accessible from the Mt. Edith Cavell parking lot. Beyond Mt. Kerkeslin the Simpson Pass Thrust approximately parallels the course of the Icefields Parkway for a distance of more than 320 km.

Photo by W.R. Price

THE WHISTLERS / 143

This view is to the west from the upper tramway terminal atop The Whistlers. The valley of the Miette River provides highway and rail access to Yellowhead Pass and the interior of British Columbia. The low, rounded hills of the valley are carved from gritstone and slate of the Precambrian Miette Group. In the far distance Cambrian quartzite and carbonate strata of the Gog Group form the Victoria Cross Ranges.

Photo by W.R. Price

Close to the upper terminal of The Whistlers tramway, quartzite strata of the Precambrian Miette Group have been folded into a small anticline.

Photo by W.R. Price

trail levels out are erratics, carried here by glacial ice from Gog Group outcrops to the west or south. They show how thick the glaciers must have been when they occupied the Athabasca and Miette valleys.

From any convenient vantage point near the summit you can see a broad panorama extending through an arc of about 180°, from the southeast along the Athabasca Valley to the northwest along the valley of the Miette River. To the southeast the prominent symmetrical mountain on the east side of the Athabasca Valley is Mt. Kerkeslin, where strata of Cambrian age have been warped into a gentle syncline. The Cambrian strata include two thin limestone formations that appear as prominent cliffs in the photograph on page 92. The lower of the two, the Mural Formation, contains abundant trilobites and coral-like creatures called archaeocyathids, which Bill Fritz of the Geological Survey has used to date these rocks as Early Cambrian, about 570 to 550 million years old.

In the discussion of Marmot Mountain we referred to the Simpson Pass Thrust as one of the major structural features of the Rockies. This thrust fault more or less follows the course of the Icefields Parkway from south of Lake Louise in Banff National Park to the Miette River west of Jasper, a distance of some 320 km. Along its length, rocks above the fault, mainly of Precambrian age, have been shoved eastward over Paleozoic rocks beneath the fault.

From your viewpoint atop The Whistlers you can see Marmot Mountain to the south (on the west side of the Athabasca Valley) and, behind it, the westerly inclined Cambrian strata of Mt. Edith Cavell. The Cavell Meadows, carved from Precambrian Miette Group rock, surround the foot of the peak's sharp eastern ridge. In the far distance are the peaks of the Fryatt Group, their summits glacially sculpted from Cambrian carbonates. On a clear day you can see the peak of Mt. Robson, 85 km to the west.

(38) Mt. Edith Cavell

Take the Cavell Road 12 km from its junction with Highway 93A. When you step out of your car into the cool, fragrant subalpine forest at the end of the Cavell Road, you know immediately that you're in a special place. The great north face of Mt. Edith Cavell rises 1380 m from base to top, all of it Cambrian Gog Group quartzite (see photo on page 97). The Angel Glacier drapes itself picturesquely over cliffs of the same strata to the right, and to the left of the face lies the flowery green tundra of the Cavell Meadows, underlain by Miette Group carbonates and slate.

A walk up the trail into the Cavell Meadows puts the icing on a day's cake. From the parking lot, find the stairs to the top of the Little Ice Age moraine and follow the paved trail to the left. The trail crosses Teahouse Creek (named for the old teahouse that used to occupy the parking lot) and continues up through the boulders of one of the Cavell Glacier's lateral moraines. It then levels off, going just beyond the moraine, where the contrast between the untouched forest and the chaotic glacial debris is striking. The trail switchbacks up through thinning woods to a well-placed observation/lunch stop, the perfect place for a photo of the Angel Glacier. From here you can see that the tongue of ice forming the angel's robe issues from a large, ice-filled cirque that can't be seen from the parking lot. The angel has lost a good bit of her robe since the turn of the century; you can get an idea of how much by noting the change in colour of the rocks on either side of the ice tongue. The rocks are lighter where the ice lay most recently, because slow-growing lichens have not yet colonized and darkened the exposed rock surfaces.

A tongue of ice forming the robe of the Angel Glacier flows out of an ice-filled cirque carved into Cambrian quartzite on the face of Mt. Edith Cavell.

Photo by Ben Gadd

Above the timberline it's possible to see the geology of the mountain clearly. The meadows are in Precambrian strata of the slaty upper Miette Group. The base of the north face follows the contact between the top of the Miette and the base of the Cambrian Gog Group. The entire face is carved in a great block of Gog quartzite, a rock type that resists erosion much better than rocks of the Miette Group. This explains why the peak stands so much higher than the surrounding landscape. Note the gentle southwesterly inclination of the strata, typical of the eastern main ranges.

At km 3.8 the trail reaches its highest point and begins to loop back down through the meadows. If the weather is good, and if you're feeling fit, you might want to take the smaller trail that branches off here and leads some 300 m higher to a minor summit (elevation 2562 m) near the foot of Cavell's lovely east ridge. That summit is made of grey Precambrian dolomite, which locally occurs near the top of the Miette Group.

146 / OTHER LOCALITIES OF GEOLOGICAL INTEREST

Cia Gadd gazes out across the Cavell Alplands, formed upon Precambrian Miette Group strata in front of Mt. Edith Cavell.
Photo by Ben Gadd

From the minor summit you can see the expansive Cavell Alplands to the north, hundreds of hectares of tundra that are home to caribou, marmots and grizzly bears. To the northeast, across the Athabasca Valley, lies the Maligne Range, also composed of Miette Group rocks; in the far distance are the grey limestone battlements of the front ranges. To the east you can see the rugged Gog quartzite slope of Mt. Hardisty and the gentle syncline in the face of Mt. Kerkeslin, also carved mostly from Gog Group strata but with carbonates of the Peyto, Snake Indian, and Titkana formations at the top. Let your eyes rove southeastward up the valley of the Sunwapta River, along the line of quartzite slabs marking Endless Chain Ridge; south of that lies Sunwapta Peak, a writing-desk-shaped mountain with a classic dip slope on one side and a glacier on the other.

On the return leg of your hike, don't miss the turnoff to Cavell Pond. It's only half a kilometre from the start of the trail and leads gently down through the bouldery glacial forefield to the grey-green, rock-flour–rich meltwater lake lying against the ice-front of the Cavell Glacier. The stone-covered ice hiding from the sun under Mt. Edith Cavell's north face is all that remains of the glacier that once covered the ground between you and the parking lot. If not for avalanches of snow and ice that slide down the cliffs to collect at the cliff base, the Cavell Glacier would not be there at all. With luck you will be able to hear the boom of such an avalanche.

The terminal moraine formed at the close of the Cavell glacial advance during the Little Ice Age provides a natural elevated pathway for visitors wishing to explore the surroundings of Mt. Edith Cavell.

Photo by Ben Gadd

Brian Luckman, who has chronicled the retreat of the Athabasca Glacier (see pages 128–30), has also studied the retreat history of the Cavell Glacier. He used a species of green-and-black lichen called *Rhizocarpon geographicum*, which adorns the quartzite blocks here. Like most lichens, this one grows slowly and at a known rate. By measuring the sizes of lichen colonies on the rocks and comparing them with tree-ring data from small Engelmann spruces sprouting among the forefield boulders, Brian was able to calibrate his lichens and work out the melt-back sequence with astonishing accuracy. As you follow the trail back to the parking lot, note how the trees get taller.

The green meltwaters of Cavell Pond beneath Cavell Glacier are characteristic of the rock-flour–rich lakes of the Canadian Rockies.

Photo by Ben Gadd

The Tonquin Valley via Astoria River and Maccarib Pass
Walking distance 43 km (27 miles)

The Tonquin Valley hike is long, but worth the effort of every step. Give yourself two or three days for this backpacking trip. The views are glorious and the geology interesting.

Most hikers do this loop in a clockwise direction, going in on the Astoria River trail, which begins near the end of the Cavell Road, hiking the length of the Tonquin Valley, and exiting over Maccarib Pass. Parks Canada has provided several back-country campsites along the route, one of them in the heart of the Tonquin Valley beside the Amethyst Lakes.

A long, easy traverse of the slopes of Sorrow Peak, made of the same block of Cambrian Gog quartzite from which Mt. Edith Cavell was carved, leads across the Moose Pass Thrust half a kilometre before you reach the bridge over the Astoria River. West of the thrust lies a strip of soft Miette Group slate, into which Verdant Pass (the valley coming in from the south) has been cut. At the bridge (5.0 km) you cross the Whirlpool Fault, a normal fault along which the west block dropped downward relative to the east block. This fault brings Gog Group strata back into the landscape in the form of the imposing quartzite mass of Throne Mountain (on your left), with its enormous northeast-facing glacial cirque.

Dark grey and rusty red-stained quartzite strata of the Cambrian Gog Group form Throne Mountain above the Astoria River valley.
Photo by C.J. Yorath

Another 3.5 km of easy walking along the river brings you to the long set of switchbacks up the north valley wall, past rockslides from the slopes of Oldhorn Mountain, the sharp peak on your right. Oldhorn is an isolated block of Gog quartzite resting upon a bed of crumpled, unstable, deeply eroded Miette Group shale. As the edges of the quartzite block tilt outward, cracks open along joints, and pillars of rock come tumbling down to collect in **talus** heaps at the base of the peak. Ice has invaded these talus heaps, turning them into rock glaciers that have crept slowly downhill.

The trail levels out at km 13, and soon you get what is arguably the finest view in the Canadian Rockies, the great quartzite wall of the Ramparts (lower photo, page 98), 13 km long and rising an average of one vertical kilometre above the Tonquin Valley (elevation 1960 m at the Amethyst Lakes). The maximum elevation along the ridge is 3322 m at Simon Peak, and the vertical relief above the lakes is 1350 m. Again, it is differential erosion that accounts for the local relief: the Ramparts are made of hard Gog quartzite, while the Tonquin Valley is cut into poorly resistant Miette Group slate.

If you planned three days for your trip, you could easily spend one of them exploring the geology of the Tonquin Valley. The Amethyst Lakes, which are actually just one lake with a very narrow connection, occupy a 4 km long hollow in glacial debris spread over the underlying impermeable slate. This basin may have been created by the melting of large blocks of glacial ice stranded here at the end of the late Wisconsinan glaciation some 11,000 years ago. Along the western shore of the lakes you can see the jumbled moraines left by Little Ice Age glaciers that advanced to the lakes between about A.D. 1200 and the mid-1840s. The ice has since retreated closer to the cliffs, where several modern glaciers are dutifully scratching away, trying to deepen their cirques in the Ramparts wall.

Halfway up the Ramparts there is a conspicuous snow-holding ledge. This is a soft layer of sandstone in the extremely hard Gog Group quartzite. Sandstone is similar to quartzite but lacks the quartz cement between the sand grains, which makes it softer and thus easier to erode. Other long ledges on the Ramparts also mark softer layers in the formation. Weathering of the Gog's ever-present iron pyrite has been more pronounced here, giving such ledges the bright rusty-orange-to-pink colours of the iron-oxide minerals **goethite** and **hematite**.

North of the Amethyst Lakes the trail leaves the Tonquin Valley and follows the gentle watershed of Maccarib Creek. This valley has been cut along the trace of the Whirlpool Fault, the same normal fault you crossed at the Astoria River bridge near the beginning of the hike. Here it separates slate from slate, and thus erosion has accented neither one side nor the other. Mt. Clitheroe, seen on the right (south) side of the valley, is

The winter slopes of Maccarib Pass are a paradise for cross-country skiers.

Photo by Ben Gadd

analogous to Oldhorn Mountain: it's another quartzite block falling to pieces over a bed of slate. In this case, the pieces have formed a rock glacier nearly 3 km wide.

Maccarib Pass is a broad gap between shaly Mt. Maccarib on the south and the quartzitic peaks of the Trident Range on the north. The slopes around the pass are renowned for two things. One is the presence of caribou, as might be expected from the name of the pass — "maccarib" is an aboriginal word for caribou. The other is the dazzling accumulation of deep powder snow in winter, giving this place a reputation among cross-country skiers for excellent telemarking.

Three kilometres northeast of the pass you again cross the Moose Pass Thrust. On the far side lies the Gog Group, where the peaks take on the Gog's craggy quartzite character. Looking down the valley through the Portal, a fitting name for the glacially eroded gap between Peveril Peak on the left and Lectern Peak on the right, you can see the back side of Marmot Mountain; on its front side, out of sight, is the ski area described on page 137. At this point the trail descends steeply to Portal Creek and follows it down through deposits of glacial debris to the trailhead at Marmot Road, the end of both this hike and this book.

Glossary

Alluvial fan. An outward spreading, fan-shaped mass of loose rock material deposited by (1), a stream where it leaves a narrow mountain valley and flows across an adjacent plain, or (2), where a tributary stream of steep gradient joins a main stream of lower gradient.

Ammonites. Many genera of fossil cephalopods with the form of a flat, coiled spiral that contained a squid-like creature.

Antecedent stream. A stream established prior to uplift of the ranges through which it has cut its course.

Anticline. An up-fold in which the strata have been bent concave downward. The oldest strata are found in the centre of the fold.

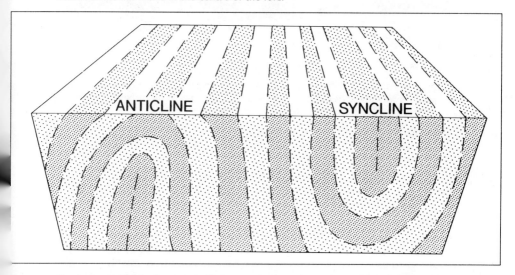

Archaeocyathids. A cone-, goblet- or vase-shaped organism with a skeleton of calcium carbonate. Archaeocyathids lived in reef colonies throughout the world's oceans during the early part of the Cambrian Period.

Arête. A narrow, serrated mountain ridge developed by the glacial sculpting of the back walls (headwalls) of adjacent **cirques.**

Atoll. A roughly circular, or horseshoe-shaped, coral reef, enclosing a shallow lagoon and commonly constructed upon the tops of individual volcanoes of **volcanic island arcs.**

Bear-pole. A bar, elevated about 4.5 m (15 feet) to 6.0 m (20 feet) above the ground and supported by two trees or poles from which one can suspend food, camping gear or unruly children out of the reach of wildlife.

Belemnites. Cigar-shaped shells of fossil cephalopods.

Brachiopods. A group of nearly extinct bivalve organisms that reached their peak in numbers and diversity during the Paleozoic Era.

Braided stream. One flowing in several dividing and reuniting channels resembling the strands of a braid. Such streams occur where more sediment is being brought to any part of a stream than it can remove. The building of bars of sediment becomes excessive, forcing the stream to develop an intricate network of interlacing channels.

Breccia. A rock that has been broken up into angular fragments, each fragment surrounded by material of similar or different composition.

Bryozoans. A group of small, aquatic animals, usually found in colonies of delicately branching structures.

Calcite. Calcium carbonate ($CaCO_3$), the mineral that forms limestone.

Canadian Shield. The vast, shield-shaped region of exposed Precambrian crystalline rocks centred upon Hudson Bay. The region underlies more than 5,000,000 km^2 and is composed of welded fragments of crust formed between 4.0 and 2.5 billion years ago.

Carbonates. As used herein, carbonates are sedimentary rocks formed by the organic precipitation from marine waters of mineral carbonates of calcium and magnesium. Examples are **limestone** (= calcium carbonate = the mineral **calcite**) and **dolomite** (= calcium and magnesium carbonate).

Cavell Advance. See **Little Ice Age.**

Chert. A dense sedimentary rock consisting of extremely fine, interlocking crystals of quartz, occurring as nodules or beds within carbonates or as extensive layers resulting from the accumulation of **radiolaria** shells.

Chlorite. A hydrous greenish mineral, resembling mica, consisting of various combinations of magnesium, iron, aluminum and silica with the general formula of $(Mg, Fe^{+2}, Fe^{+3})_6 AlSi_3O_{10}(OH)_8$.

Cirque. A steep-walled, amphitheatre-like recess occurring at high elevations on the side of a mountain, commonly at the head of a glacial valley and formed from a mountain glacier's erosive carving. **Tarn lakes** commonly occur at the bases of cirque headwalls.

Clastics. Sedimentary rocks composed of broken fragments (= **clasts**) derived through erosion of pre-existing rocks or minerals and transported by streams, wind, etc., to their place of deposition. The most common clastic rocks are quartz sandstone and shale.

Clay. Rocks or mineral fragments of any composition having diameters of less than 1/256 mm. The mineral fragments commonly belong to a complex and loosely defined group of hydrous silicates of aluminum, iron and magnesium.

Cleavage. The tendency of some rocks to cleave or split along parallel, closely spaced fracture planes that may be parallel to or obliquely inclined to the stratification. Cleavage is developed under conditions of increased pressure, commonly during episodes of mountain-building.

Columbia Mountains. Occurring in southeastern British Columbia, the Columbia Mountains include the Cariboo, Selkirk and Purcell mountains.

Competence. (1) The ability of a stream to transport a given sediment load; and (2) the ability of rocks, depending upon their strength, rigidity and composition, to withstand deforming erosive forces.

Conglomerate. A coarse-grained clastic sedimentary rock, composed of rounded to subangular fragments larger than 2 mm in diameter within a matrix of sand or silt (= consolidated gravel).

Continental divide. A drainage divide separating streams flowing toward separate oceans. In the Jasper Rockies the Continental Divide separates streams flowing into the Arctic, Pacific and Atlantic (Hudson Bay) oceans.

Cordillera. The system of mountain ranges, valleys and plateaus extending from the southern tip of South America to the Arctic Ocean. It forms the backbone, or continental spine, of western North America.

Cordilleran Ice Sheet. See **Wisconsinan glaciation.**

Cornice. An overhanging ledge of snow on the edge of a steep ridge or cliff face.

Cross-bedding. A series of internal sedimentary layers, inclined at low angles to the main stratification. Similar to herringbone tweed.

Crown. The rock remaining in place adjacent to the highest parts of a scarp (steep slope) from which a landslide moved.

Crust. The earth's crust is of two types: (1) the crust that forms the continents is between 32 and about 55 km thick and composed mainly of minerals of sodium, potassium, aluminum, silicon and oxygen; and (2) the oceanic crust, excluding its thin cover of sediments, is about 7 km thick and dominantly made of compounds of iron, magnesium, silicon and oxygen.

Debris flow. A water-saturated mass of loose soil and coarse debris moving downslope.

Décollement. As used in this guide, the principal surface from which the sedimentary rocks forming the Rocky Mountains were detached and moved eastwards over the Precambrian crystalline granitic rocks of the buried western flank of the Canadian Shield. The décollement forms a common zone wherein the several main thrust faults merge into a common surface of detachment.

Diamictites. As used in this guide, a rock containing a large variety of different types and sizes of clasts in a matrix of clay or mudstone and deposited upon the seafloor through the melting of glaciers or ice shelves.

Differential erosion. Erosion that occurs at varying rates because of differences in the resistance of adjacent rock types.

Dip. The angle of downward inclination from the horizontal of a planar surface such as a bedding surface, fault, etc. **Dip slope** is a surface slope parallel to and formed by the dip of strata.

Dolomite. See **Carbonates.**

Dolomitization. The conversion of limestone to dolomite through the addition of magnesium. Dolomitization occurs shortly after deposition of limestone through the action of magnesium-bearing marine water.

End moraine. A ridge of unsorted gravel, sand and silt deposited at the snout of an active glacier (see **Terminal moraine**).

Erratic. As used in this guide, a large boulder that has been carried to its resting place by glacial ice. Far-travelled erratics are usually recognized by their difference in composition from the bedrock upon which they rest.

Feldspars. A group of closely related tabular minerals composed of silica, oxygen, aluminum, and one or more of potassium, sodium and calcium. The feldspars are the most abundant of all minerals in the earth's crust and are formed by crystallization from molten **magmas.** They are the most common constituents of igneous rocks and occur as mineral grains in sedimentary rocks.

Folds. See **Anticline** and **Syncline.**

Forefield. A general term embracing the melt-back area of a glacier bounded by the end and lateral moraines and the ice-front.

Formation. See **Group.**

Foothills. The region of low, rounded hills fringing the easternmost range of the Rockies and bounded to the east by the flat to gently rolling plains.

Front ranges. Those northwesterly trending ranges occurring between the western limit of the foothills and the eastern edge of the main ranges. The front ranges consist mainly of Paleozoic and younger carbonate strata that occur in westerly inclined thick thrust sheets, each range separated by linear valleys underlain by shales.

Garnets. A group of minerals consisting of various combinations of calcium, magnesium, iron, aluminum, manganese, vanadium, chromium, silica and oxygen. Although they occur as minor constituents of igneous rocks, they are most commonly found as distinctive red crystals in metamorphic rocks, in which they formed at high pressures and temperatures.

Geological Survey of Canada (GSC). Canada's largest scientific research organization and one of the older and more prestigious of its kind in the world. Since its establishment in 1842 the GSC has provided Canadians with an understanding of the geological architecture of the country with a view to fostering the responsible development and use of Canada's mineral and energy resources as well as a knowledge of its natural hazards. With headquarters in Ottawa, it has regional offices in Dartmouth, Nova Scotia; Quebec City; Calgary, Alberta; and Vancouver and Victoria, British Columbia.

Glacier. A large mass of long-lasting ice that is formed, on land, by the recrystallization and compaction of snow and that moves slowly downslope or outward in all directions because of the stress caused by own weight.

Glaciation. The formation, movement and recession of ice sheets. A collective term for the geological processes associated with glacial activity, including erosion and deposition, and the resulting effects of such action on the earth's surface.

Gneiss. See **Metamorphic rocks.**

Goethite. A hydrous iron oxide occurring as a common constituent of natural rust.

Gondwanaland. See **Pangea.**

Graded beds. Beds of sedimentary and some volcanic strata in which the size of clasts changes in a regular fashion from the base of each bed to the top.

Graptolites. Tiny colonial floating organisms that lived throughout the Paleozoic Era. The organisms consisted of a rigid "spine" along which were arranged individual living chambers occupied by a polyp-like organism. Graptolites are commonly found on the bedding surfaces of black shales, where they resemble narrow leaves.

Grit, gritstone. As used in this guide, a coarse-grained sandstone composed of angular fragments of quartz, feldspar, other minerals and rock fragments.

Ground moraine. An accumulation of debris released from a glacier during melt-back and forming an extensive area of low relief.

Group. Stratified rocks are given names according to many criteria, including their composition and location. The most fundamental type of subdivision is the **formation,** which refers to a succession of strata with specific characteristics and which can be recognized as distinctly different from other formations throughout its region of occurrence. For example, the **Old Fort Point Formation** is the name given to a succession of multi-coloured slates that occur widely throughout the Jasper Rockies but are best exposed at Old Fort Point, where they were first studied. In some instances several formations sharing common origins or other characteristics can collectively be referred to as a **group**; the Old Fort Point Formation is one of several formations assigned to the **Miette Group**. There are **supergroups**, too. The Miette Group is part of the **Windermere Supergroup**.

Hanging valley. A tributary glacial valley, the mouth of which is at a relatively high elevation on the steep side of a larger glacial main valley.

Hematite. Iron oxide.

Horn peak. A high, rocky, sharp-pointed mountain peak with prominent faces and ridges, bounded by the intersecting walls of three or more **cirques** that have been cut by the headward erosion of mountain glaciers.

Icefall. The accumulation of ice resulting from the breaking away of a glacier as it flows over the edge of a steep cliff.

Icefield. An extensive mass of land ice covering a mountain region, consisting of many interconnected alpine glaciers and covering all but the highest peaks and ridges.

Ice-contact stratified drift. Glacial drift that accumulates upon or against the sides of a melting glacier. As the ice seasonally moves, soil and rock materials entrained in the ice accumulate in layers, usually in low points on the ice surface or in water that has collected in ponds against its sides.

Insular Superterrane. See **Superterranes.**

Intermontane Superterrane. See **Superterranes.**

Isostasy. A condition analogous to floating whereby when a load is placed upon the earth's crust, the crust sinks proportionally to the magnitude of the applied load, and when a load is removed from the earth's crust, the crust rebounds porportionally.

Joints. Fractures, commonly occurring in two or more sets, on either side of which there has been no appreciable movement of the rocks.

Karst. A type of topography that is formed by the dissolution of rock in water, characterized by sinkholes, caves or underground drainage. Most karst develops in regions of limestone.

Kettle. A steep-sided, usually basin- or bowl-shaped depression in glacial drift deposits, often containing a lake and formed by the melting of a large, detached block of ice left behind by a retreating glacier.

Klippe (plural = **klippen**). As used in this guide, an isolated, thrust-faulted sheet of strata that has been separated from the remainder of the sheet by erosion.

Laurentide Ice Sheet. See **Wisconsinan glaciation.**

Laurasia. See **Pangea.**

Limestone. See **Carbonates.**

Lithification. The processes by which sediments such as gravel, sand and mud are converted to rocks such as conglomerate, sandstone and shale.

Little Ice Age. A period when mountain glaciers attained their maximum extent during the past several centuries. In the Rockies glaciers began their advance

(= **Cavell Advance**) about 950 years ago and reached their maximum extents between 100 and 400 years ago.

Loess. A loose deposit of silt and clay deposited by wind.

Magma. Molten rock material from which igneous rocks and minerals form by cooling and solidification. Most magmas form by partial melting of the mantle, from where they are injected into the overlying crust.

Main ranges. Those ranges of the Canadian Rockies along the Continental Divide consisting mainly of Cambrian and Precambrian strata, the latter deformed into complex folds and thrust faults. Cambrian rocks consist of thick, undeformed quartz sandstones and carbonates. The main ranges are where the greatest rise occurred when the Rockies were formed.

Mantle. The thick layer between the base of the earth's crust and the top of the core, composed largely of compounds of iron, magnesium, silicon and oxygen. Parts of the upper mantle are soft, since high pressures and heat partially melt the rocks in this region.

Metamorphic rocks. Rocks which, since their initial formation, have sustained sufficient increase in temperature and pressure such that their original minerals and textures have been changed to new minerals and new textures. **Slate** is the metamorphic equivalent of **shale** and can be split into thin slabs. **Schist** is a foliated, crystalline rock that can be split into slabs because of the parallelism of the minerals present. **Gneiss** is a foliated rock that commonly has bands of alternating layers of dark- and light-coloured minerals. It does not readily split into slabs.

Normal fault. A steep fault, usually inclined at an angle greater than 45°, along which the rocks above the fault have moved downward relative to those beneath the fault. Whereas **thrust faults** result from compressive forces, normal faults occur when the rocks are subjected to tensile (stretching) forces.

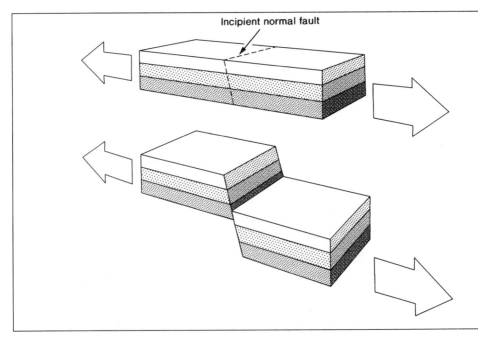

GLOSSARY / 159

Nunatak. An isolated knob or peak of bedrock that projects prominently above the surface of a surrounding glacier.

Obduction. The overthrusting of oceanic crust onto the edges of continents.

Omineca Belt. One of five subdivisions of the Canadian Cordillera, differentiated from one another by a combination of rock types, geological history and physiography. The easternmost belt, containing the Rockies, is called the **Foreland Belt,** consisting entirely of sedimentary rocks. To the west, the Omineca Belt, containing the Purcell, Selkirk, Monashee, Cariboo, Omineca, Cassiar and Selwyn mountains, consists of sedimentary, igneous and metamorphic rocks. The **Intermontane Belt** in the central part of the Cordillera is dominated by volcanic rocks and forms the broad interior plateaus of British Columbia and the Yukon, but it also contains the rugged Skeena Mountains of north-central British Columbia. To the west is the **Coast Belt,** consisting of rugged peaks of granitic igneous rocks, and to the west again, the **Insular Belt,** formed of sedimentary and igneous rocks underlying Vancouver Island, the Queen Charlotte Islands and the Saint Elias Mountains, the highest and most ruggedly beautiful in North America.

Pangea (also **Pangaea**). The name given to the giant supercontinent (made up of all of the world's continents) that had assembled by about 300 million years ago and began to break up about 200 million years ago. The initial breakup resulted in the separation of **Laurasia,** including North America and Eurasia, from **Gondwanaland,** which contained South America, Africa, India, Australia and Antarctica.

Panthalasa. The globe-encircling ocean surrounding **Pangea.**

Patterned ground. As used in this guide, polygons of coarse fragments surrounding finer particles resulting from the alternating freezing and thawing of **permafrost** in arctic and alpine regions. If the surface is sloping, the polygons stretch out downhill to become **stone stripes.**

Pediment. A broad, gently sloping, bedrock-floored erosion surface developed in desert regions at the base of a mountain front.

Permafrost. A soil or other surface deposit, occurring in arctic or alpine regions, that has been permanently frozen for several years. During the summer, its upper layer (= active layer) will thaw.

Physiography. The description of the earth's surface features and landforms.

Plate tectonics. A corollary of **seafloor spreading,** plate tectonics explains the motions of the world's crustal plates and the **subduction** of old oceanic crust beneath the world's deep sea trenches.

Pothole. As used in this guide, a pot-shaped pit or hole eroded into bedrock by the erosive action of stones moved in a swirling fashion by fast-moving water.

Precambrian Eon. That period of earth history between its formation, some 4.6 billion years ago, and about 570 million years ago when life increased dramatically in both numbers and diversity. Strictly speaking, our use of this term is incorrect. Precambrian time is divided into two eons, the **Archean Eon,** ending about 2500 million years ago, and the **Proterozoic Eon.** For simplicity we have chosen to employ the term "Precambrian Eon" and reduce "Archean" and "Proterozoic" to era status, neither of which we use in the text. Our apologies to purists.

Quartz. A mineral compound of silica and oxygen, most commonly colourless or white, originally formed during crystallization of molten magma and a common constituent of igneous, metamorphic and clastic sedimentary rocks.

Quartzite. A hard, unmetamorphosed sandstone, consisting chiefly of quartz grains cemented together by secondary crystalline quartz.

Radiolarian ooze. Radiolaria are planktonic, single-celled protozoan animals, the skeletons of which are made of silica. The constant rain of these tiny skeletons upon the deep seafloor results in the accumulation of oozes.

Recessional moraine. An end or lateral moraine constructed during a temporary but significant pause in the final retreat of a glacier.

Ripple marks. Parallel, small-scale ridges and hollows produced by currents acting upon unconsolidated sediments such as those seen on modern beaches where waves are capable of moving grains of sand.

Rock flour. As used in this guide: fine powder formed when stones embedded in a glacier grind down the underlying bedrock.

Rock glacier. A lobate or tongue-shaped, slowly moving mass of poorly sorted angular boulders and pebbles cemented together by ice.

Rocky Mountain Trench. Part of a system of linear, deep, fault-controlled valleys extending the entire length of the Canadian Cordillera.

Rodinia. The name given to a giant supercontinent incorporating all of the world's continents, formed about 1750 million years ago. Rodinia broke up about 750 million years ago.

Sandstone. A clastic sedimentary rock composed of rounded to angular fragments of sand-size, set within a matrix of silt or clay, and cemented by calcite, silica or iron oxides. The sand particles most commonly consist of quartz; however, other minerals such as feldspars and micas commonly are included.

Schist. See **Metamorphic rocks.**

Scour marks. A groove produced by the cutting or scouring action of currents moving across unconsolidated sediments.

Scree. Small pieces of rock that have weathered from a cliff and accumulated at the base. See also **Talus.**

Seafloor spreading. The process that explains the creation of oceanic crust by convective upwelling of molten magma along the global mid-ocean **spreading ridge** system and the growth, or movement away from the ridge system, of the newly forming crust.

Seamount. As used in this guide, a volcanic cone constructed upon the seafloor. Upon reaching the surface a seamount becomes a volcanic island.

Shale. A fine-grained clastic sedimentary rock formed from the consolidation of clay and silt (= lithified mud).

Slate. See **Metamorphic rocks.**

Solifluction lobe. A slow-moving, isolated, steep-fronted, tongue-shaped, water-saturated mass of soil and other material. The lobes are commonly covered with a mat of vegetation that retards their speed of motion and contributes to their steep fronts.

Splay fault. A thrust fault that splays off from another thrust fault.

Spreading ridge. See **Seafloor spreading.**

Stone stripes. See **Patterned ground.**

Strata. Plural of **stratum.** Layers of sedimentary or volcanic rocks, either horizontal, tilted or otherwise deformed.

Striations. As used in this guide, linear scratches or grooves carved into bedrock by stones embedded into the base of a moving glacier.

Stromatolites. Primitive organic structures created by algae as they trap sediment or precipitate calcium carbonate to form tabular, domelike or bulbous shapes in limestone.

Stromatoporoids. Extinct marine, reef-building organisms of uncertain biological affinity. They secreted skeletons of calcium carbonate in a wide variety of shapes and were especially abundant from Ordovician through Devonian time.

Structural control. As used in this guide, the influence of structural features on the development of landforms.

Structural style. A general term describing the manner by which rocks of a given region have been deformed. For example, the structural style of the southern Rockies is one dominated by thrust faults, whereas the structural style of the Mackenzie Mountains is expressed mainly by folds.

Subduction. The process of consumption of oceanic crust at the deep-sea trenches whereby one piece of crust descends, or is **subducted,** beneath another. An example occurs off our west coast where the oceanic rocks of the Juan de Fuca Plate, created at the Juan de Fuca Ridge, have spread away from the ridge through **seafloor spreading,** and are being consumed along the Cascadia subduction zone off the west coast of Vancouver Island, Washington, Oregon and northernmost California.

Subsequent stream. A tributary stream formed after uplift of the mountains and which flows parallel to the trends of the ranges.

Superterranes. Assemblies of smaller terranes that amalgamated prior to their accretion to western North America. Western Canada is constructed from two such superterranes. The **Intermontane Superterrane** consists of several smaller terranes with names such as "Stikinia," "Quesnellia," "Cache Creek" and "Slide Mountain." These smaller terranes amalgamated about 225 million years ago and collided with North America about 170 million years ago. The **Insular Superterrane,** consisting of smaller crustal fragments called "Wrangellia" and the "Alexander Terrane," crashed into the previously accreted Intermontane Superterrane about 100 million years ago.

Surface trace of a fault. The line of intersection of the plane of a fault with the earth's surface (= fault line).

Syncline. A down-fold in which the strata have been bent concave upward. The youngest strata are found in the centre of the fold.

Talus. Rock fragments of any size or shape, but especially large, bouldery pieces, eroded from and lying at the base of a cliff or steep rocky slope. Talus slopes consist of angular rock debris formed by rockfalls and the continual spalling of material from higher cliffs.

Tarn lake. See **Cirque.**

Tectonic. An adjective referring to large-scale rock structure and the deformational history of the earth's crust in a given region.

Terminal moraine. The end moraine that marks the farthest advance or maximum extent of a glacier.

Terranes. Parts of the earth's crust that preserve geological records different from those of neighbouring parts. The boundaries between terranes are faults. Some appear to be comparatively thin sheets, whereas others are at least 18 to 20 km thick. Some terranes originated close to their present position, while others were formed thousands of kilometres from their current location.

Thrust fault. A fault surface, generally inclined at less than 45°, along which the rocks above the fault have moved upward and over top of the strata below the fault. In the Canadian Rockies thrust faults separate thick sheets of strata (= **thrust sheets**) that have been shoved eastwards many tens to hundreds of kilometres. In most cases in the Canadian Rockies, the rocks immediately above the fault surface are older than those immediately below.

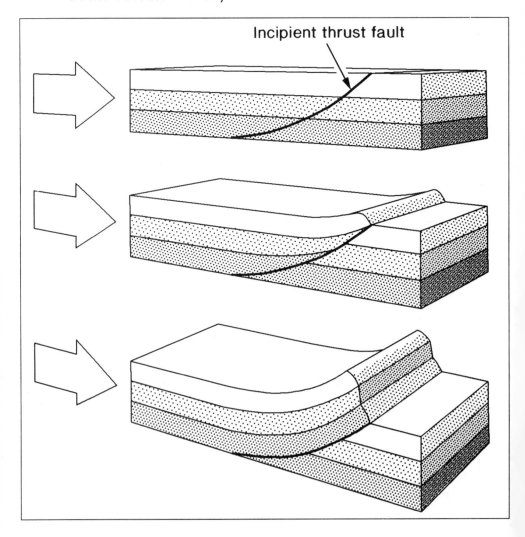

Trellis drainage. A drainage pattern characterized by parallel main streams intersected at, or nearly at, right angles by their tributaries, which, in turn, are fed by secondary tributaries flowing parallel to the main streams.

Trilobites. A segmented, three-lobed, bottom-dwelling marine arthropod that lived during the Paleozoic Era, most abundantly during the Cambrian and Ordovician periods. Our favourite is *Ogygopsis klotsi*.

Turbidites. Sediments deposited from dense, sediment-laden currents. Turbidites commonly result from submarine landslides and are recognized by specific kinds of internal sedimentary structures, such as graded beds and cross-laminations resulting from the ripple-producing flow of the current.

Unconformity. As used in this guide, a gap in the geological record (= absence of strata) between two formations whereby the upper formation is substantially younger than the lower, thus indicating that an interval of earth history spanning the age limits between the two formations is locally unrepresented by rock strata.

Volcanic arcs. Chains of volcanoes constructed above subducting oceanic crust. As oceanic crust is subducted, it reaches depths where the temperature is sufficient to melt the subducting plate; molten material rises upward from this region through the overriding plate to appear at the surface as a chain of volcanoes. In the oceans these chains are called **volcanic island arcs** and on land they are called **volcanic arcs**.

Wisconsinan glaciation. The last of several major advances of continental and Cordilleran ice sheets. These glaciers covered much of the Northern Hemisphere between 80,000 and 10,000 years ago. Those affecting North America were called the **Laurentide Ice Sheet** and the **Cordilleran Ice Sheet**.

Sources and Additional Reading

Baird, D.M. 1963. Jasper National Park – Behind the mountains and glaciers. Geological Survey of Canada, Miscellaneous Report No. 6.
Although out of date, this book remains a useful reference.

Cruden, D.M. 1976. Major rock slides in the Rockies. *Canadian Geotechnical Journal* 13:8–20.
A discussion of the sizes and causes of several large landslides that have occurred in the Rockies; no estimates of when.

Gabrielse, H., and C.J. Yorath, eds. 1991. *Geology of the Cordilleran Orogen in Canada.* Geological Survey of Canada, Geology of Canada, no. 4, 844 p.
A comprehensive, technical account of the geological architecture, tectonic history and economic mineral and energy resources of the Canadian Cordillera.

Gadd, Ben. 1995. *Handbook of the Canadian Rockies.* Jasper, Alta: Corax Press. 878 p.
A compilation of geological, botanical, zoological, historical and recreational features of the Canadian Rockies.

Mountjoy, E.W. 1980. Geology, Mount Robson. Geological Survey of Canada, Map 1499A, scale 1:250 000.
A coloured map covering the region from the Rocky Mountain Trench to the Alberta foothills in the Mount Robson Provincial Park/northern Jasper National Park areas.

Mountjoy, E.W., and R.A. Price. 1985. Geology of Jasper, Alberta. Geological Survey of Canada, Map 1611A, scale 1:50 000.
A coloured geological map of the region north, west and south of the town of Jasper.

Mountjoy, E.W., and R.A. Price. 1988. Geology, Amethyst Lakes, Alberta – British Columbia. Geological Survey of Canada, Map 1657A, scale 1:50 000.
A coloured geological map of the Amethyst Lakes region.

Mountjoy, E.W., and R.A. Price. In press. Geology, Medicine Lake, Alberta. Geological Survey of Canada, Map, scale 1:50 000.

Mountjoy, E.W., R.A. Price and D. Lebel. In press. Geology of Mountain Park. Geological Survey of Canada, Map, scale 1:50 000.

Patton, Brian, and Bart Robinson. 1990. *The Canadian Rockies Trail Guide.* Banff, Alta.: Summerthought Ltd.
The standard reference to hiking in the Canadian Rockies.

Price, R.A., H.R. Balkwill, H.A.K. Charlesworth, D.G. Cook and P.S Simony. 1972. Field excursion A15-C15. The Canadian Rockies and tectonic evolution of the southeastern Canadian Cordillera. XXIV International Geological Congress, Montreal, Que. Published by the Geological Survey of Canada.
Geologists love to show their rocks to other geologists. This is a technical guidebook for field trips in parts of the southern Rockies. (Out of print.)

Price, R.A., D.F. Stott, R.B. Campbell, E.W. Mountjoy and N.C. Ollerenshaw. 1973. Athabasca River, Alberta – British Columbia. Geological Survey of Canada, Map 1339A, scale 1:1 000 000.
A coloured geological map covering all of NTS 83 (52°N to 56°N; 112°W to 120°W) and including all of Jasper National Park.

Wheeler, J.O., R.B. Campbell, J.E. Reesor and E.W. Mountjoy, 1972. Structural style of the southern Canadian Cordillera. XXIV International Geological Congress, Montreal, Que. Published by the Geological Survey of Canada.
A technical guidebook for field trips through parts of the Rocky, Cariboo, Monashee and Selkirk mountains. (Out of print.)

Yorath, C.J. 1990. *Where Terranes Collide.* Victoria, B.C.: Orca Book Publishers.
A semi-technical, popular account of the geological architecture, tectonic history and geological people of the Canadian Cordillera.

Index

Amethyst Lakes, **98**,* 150
Ancient Wall, 12
Ancient western edge of North
　America, **9**
Angel Glacier, 144, **145**
Anticline, 39, 40, **40**, **83**, **143**
Antler Mountain, 110
Archaeocyathid reefs, 11
Arctic drainage, 28
Arctomys Cave, 51
Arctomys Formation, 37, 51, **88**, 130
　salt crystals in, **131**
Ashlar Ridge, 72–74, **80**, **87**
Athabasca Falls, **94**, 115
Athabasca Glacier, **94**, 126, 128
　end moraine, 129
　essential facts of, 129–30
　forefield, 129
　melt-back sequence, 129
　recessional moraine, 128, 129
　striations on bedrock, 128, **128**
　tiger rocks, **129**
Athabasca River, 28, 32, 60, 64, 133
Athabasca Valley, **98**
Athabasca Valley glacier, 32, 81, 99, 119, 126, 128

Bald Hills, 108
Banff Formation, 59, 72, **86**, 131, 135
Beauty Creek Flats, **122**
Bedson Ridge, 60, **86**
Berg Glacier (Tumbling Glacier), 51
Berg Lake, 51
Bison Creek Formation
　tension cracks in, **126**
Boundary between main ranges and
　front ranges, 57–58, **58**, 81
Brazeau River, 32
Brûlé Thrust Fault, 22
Bubbling Springs, 119, **120**

Cadomin Formation, 70, **71**, **72**
Cairn Formation, **15**, 78
Canadian Cordillera
　defined, ix, 1

Canadian Shield, 1–2
Carbonate reefs
　oil fields in, 12
Carbonates
　types of, 3
Cariboo Mountains, 52
Castleguard Mountain, 127
Cathedral Formation, 119
Cavell Advance (Little Ice Age), 32
Cavell Alplands, **146**
Cavell Glacier, 144, 147
Cavell Meadows, 144
Cavell Pond, 147, **148**
Chetamon Mountain, **38**
Chetamon Thrust Fault, 22, **38**, 59
Chetamon Thrust Sheet, 108
Chetang Formation, 51
Clastics
　types of, 2–3
Cliffside Cave, **27**
Cold Sulphur Spring, 62–63, **62**
Colin Range, 102
Colin Thrust Fault, 22
Columbia Icefield, 6, **94**, 126, **127**
　hydrographic apex of North
　　America, 127–28
Columbia Mountains, 52
Conglomerate, 2–3, **4**
Continental Divide, **35**, 46–47
Cordilleran Ice Sheet, 30, 133
Cottonwood Fault, 26, **134**, 135
Curator Lake, 110
Curator Mountain, 110
Cyanobacteria, 3

Diadem Creek, 121
Differential erosion, 68, 72
Dolomite, 3, **3**
Dolomitization, 5

Earthquakes
　epicentres and magnitudes in
　　Jasper National Park, 135, **136**
East Pacific Rise, 17
Eldon Formation, 37, **38**, 51, **88**, 124

---

* **Boldface** page numbers indicate photographs and illustrations.

Endless Chain Ridge, 119, **121**, 147
Erosion
　agents of, 26
　amount of, 26–30
　estimates based upon cave elevations, 27
　rates of, 26–27
Erratic, 79, **79**
Evelyn Creek, 109
Exshaw Formation, 131–32

Fernie Formation, 74–76, **75**
Fiddle River, 60
Flume Formation, 57, 59, 62, **62**, **88**
Foothills
　origin of, 22–25, **24**
Fraser River, 49

Geological map
　description of, 39–41
　geological column of Jasper National Park, ii
　geological cross-section, 39–41
Geological/physiographic subdivisions of northwestern North America, **viii**
Glacial erratic, **31**, **79**
Glacial till, **31**
Glaciation, 30–33, 54, **55**
Global tectonic plates, **18**
Gneiss, 5
Gog Group, 10, 22, 41, 47, 51, 70, 79, **79**, **84**, **93**, **96**, **97**, **98**, 103, 115, 137, **143**, 145, 149, 150, 152
Gondwanaland, 17
Greenock Thrust Fault, 22
Grisette Mountain, 59, **60**

Henry House Flats, 59, **60**
Hoff Anticline, **69**
Hota Formation, 51
Hotsprings Thrust Fault, 77, **77**, 80

Insular Superterrane, 20, **20**
Intermontane Superterrane, 19, **20**, 23, 74
Isostasy
　defined, 28, **29**
　effects of, 28
　elevation of the Rockies, 28

Jasper Lake
　dust storms, 64, **65**
　formation of, 64–66
　river channels in, **87**
　sand dunes, 64
Jasper National Park
　boundaries of, 34
　geography, xi
　trends of mountain ranges, 34
Jasper Park Lodge, 3, 103, 135, **134**
Jasper Thrust Fault, **134**, 135, 138, **140**
Jasper townsite, 133–36, **133**
　geological map of, **134**
Jonas Slide, **93**, 119, 120
Juan de Fuca Ridge, 17

Karst drainage system, 103
Karst topography, 99, 109
Kettles, 109
Kinney Lake, 51
Kitchener Flats, 124
Kitchener Slide, 124
Klippe, **40**

Lac Beauvert, 102, 134, **135**
Landscape development of the front ranges, 53–57, **54**
　influence of climate upon, 53–54
　influence of structure upon, 53
Landslides, **93**, 102–5, **104**, **105**, **107**, 119, 124
Laurentide Ice Sheet, 30, 32
Leah Peak, 108
Lectern Peak, 152
Limestone, 3, **3**
Little Ice Age, 32, 115, 128, 150
　terminal moraine, **147**
Little Shovel Pass, 109
Livingston Formation, 80
Llysyfran Peak, 108
Loess, 64
Luscar Group, 72, **73**
Lyell Formation, 123, 130
Lynx Formation, 37, 59, **63**

Maccarib Creek, 150
Maccarib Pass, **151**, 152
Main ranges, 37, 41
  divisions of, 37
  uplift of, 37
Maligne Canyon, 81–101, **82**
  origin as a river-eroded gorge, 81–99
  origin as an unroofed cave system, 99–101
Maligne Formation, **15**, **88**
Maligne Lake, 104–8, **106**
  glacial and landslide origin of, 104–5
Maligne Range, 102
Maligne River, 32, 99, 102
Maligne River Valley
  landslides, **103**, 104, **104**, **105**
Maligne Valley Glacier, 81, 99
Marmot Mountain, 26, **98**, 137, 152
Medicine Lake
  landslide origin of, 102
  seasonal lake levels of, **90**, 102
Metamorphic rocks
  types of, 5
Miette Group, **5**, 8, 10, 22, 41, **45**, 47, 51, 57, 81, **83**, **84**, 108, 110, 115, 137, 138, 142, 145, 150
  Old Fort Point Formation, 133, 138–41, **139**, **140**, **141**
  submarine landslides, turbidites, 44–46
Miette Hotsprings, 76–79, **76**
  origin of hot water, 76–77, **77**, **78**
Miette Range, 13
Miette reef complex, 13, **14**
Miette River Valley, **143**
Miette Thrust Fault, 22, 67
Mistaya Formation, 123, 124, **125**
Mona Lake, 109
Monashee Mountains, 52
Monkhead Peak, 108
Moose Lake, 49, **50**
Moose Pass Thrust Fault, 149, 152
Morris Creek, 74
Mt. Andromeda, 127
Mt. Athabasca, 127, 130, **130**, **cover**
Mt. Brazeau, **106**, 108
Mt. Charlton, 108

Mt. Clitheroe, 150
Mt. Coleman normal fault, 123, **123**, 131
Mt. Columbia, 127, **127**
Mt. Edith Cavell, 6, 11, 30, 32, 37, **97**, 144–49
Mt. Fitzwilliam, 26, 41, 49–51, **84**
Mt. Fryatt, 30, 115, 117, **117**
Mt. Hardisty, 147
Mount Hawk Formation, **15**, 59, 67, **88**
Mount Head Formation, 79
Mt. Kerkeslin, 11, **92**, 143, 147
Mt. Kitchener, 124, 127
Mt. Mary Vaux, **106**, 108
Mt. Paul, 108
Mt. Robson, 11, 41, 49–51, **85**
Mount Robson Provincial Park, 11, 44, **83**
Mt. Stutfield, 124
Mt. Tekarra, 11, **91**, **98**, 110, 114
Mt. Unwin, 108
Mt. Wilcox, 130
Mountain Creek, 70
Mountain front
  differences in physiography at, 67–69
  differential erosion, 68, 72
Mousehole Cave, 99, **100**
Mural Formation, 51, 143

Nigel Peak, 130–31, **132**
Nikanassin Range, 80
Normal faults, 25, 39
North Twin Mountain, 127
Notch (The), 110
  view from, 111, **111**

Old Fort Point, 138–41
Old Fort Point Formation, 138
  close-up views of, **139**, **140**, **141**
Oldhorn Mountain, 150
Opal Hills Slide, 105, **107**
Outram Formation, 123

Pallisade (The), 58–59, **86**
Palliser Formation, 59, 60, **61**, 62, 67, 72, 74, 81, **86**, **87**, **88**, **89**, 102, 131, 135

INDEX / 169

Pangea (supercontinent), 13,
  break-up of, 13, **16**, 17
Paskapoo Formation, 28
Patterned ground, 111, **112**
Perdrix Formation, **15**, 57, 59, 67,
  88
Permafrost, 111
Peveril Peak, 152
Peyto Formation, 147
Physiographic development of the
  Rockies, **32**
Physiographic divisions of the
  Rockies, 34–39, **36**
  boundary between, 39
  foothills, 34
  front ranges, 34–37
  main ranges, 37
Physiographic map of the Canadian
  Cordillera, xiv
Pika Formation, 37, 51, **88**, 123
Pocahontas Mine
  coal, 72, **73**
Portal (The), 152
Portal Creek, 152
Potholes, 101, **101**
Punchbowl Falls, 70, **71**
Pyramid Creek, 57
Pyramid Lake, **96**
Pyramid Mountain, 11, 22, 37, **96**,
  137–38
Pyramid Thrust Fault, 22, 30, 57, 58,
  81, 99, 102, 109, 137
Pyramid Thrust Sheet, 58

Queen Elizabeth Ranges, 103, 108,
  109

Ramparts (The), 11, **98**, 150
Ranger Canyon Formation, 103
River Rock, 60–62, **61**
Roche à Perdrix, 60, 67, **89**
Roche Bonhomme (The Old Man),
  59, **60**, **95**, 135
Roche Boule cirque, **55**
Roche Miette, 59, 67, **88**
  erratic on, **31**
Rock flour, 115–17
  mountain goats use of, **116**, 117
Rock glaciers, **91**, 109

Rocky Mountain Trench, 51–52
  Northern Rocky Mountain
    Trench, 5
  Southern Rocky Mountain
    Trench, 52
Rocky River, 32, 64, **66**
Rodinia (supercontinent), **7**
  break-up of, 8
Rosemary's Rock, 103, **104**
Rundle Group, 70, 74, 80, 131,
  135

Samson Peak, 108
Sandstone, 2
Saskatchewan Glacier, 128
Sassenach Formation, 67
Schist, 5
Seafloor spreading and plate tec-
  tonics
  mechanisms, 13, 17
Sedimentary strata
  kinds of, 2–3
Selwyn Range, 51–52
Shale, 3
Shovel Pass (Big Shovel Pass), 109,
  110
Signal Mountain
  view from, **113**, 114
Simon Peak, 150
Simpson Pass Thrust Fault, 128, 130,
  137, 144
Sinking Ship Slide, 104, **107**
Skoki Formation, 123
Skyline Trail, 108–14
  campsites along, 109
Slate, 5, **5**
Snake Indian Formation, 37, **38**, 147
Snake Indian River, 32, 64
Snaring Thrust Fault, 39, **96**, 102,
  134, 135, 137
Snow Dome, **127**, 128
Snowbowl, 109
Solifluction lobes, 113
Sorrow Peak, 149
Southesk Formation, **15**
Stone stripes, **112**, 113
Stream pattern development in the
  front ranges, 54–57, **56**
Stromatoporoid reefs, 12, **12**

Structure of the front ranges, 53
  thrust faults, 53
Stutfield Glacier, **95**, 124
Stutfield Peak, 127
Sub-Devonian unconformity, 59, 63–64, **63**
Sulphur Creek, 80
Sulphur Mountain Formation, 79, 103, 108
Sulphur Ridge, 79
  view from, **80**
Sulphur Skyline Trail, 79–80
Sunwapta Canyon, **118**, 119
Sunwapta Falls, 119
Sunwapta Lake, 124–25
Sunwapta Pass, 131
Sunwapta Peak, 147
Sunwapta River, 121–22, 124
Sunwapta Valley Glacier, 119
Superterrane collisions with North America
  effects of, 19–21, **20**, 25–27
Survey Peak Formation, 59, 123
Syncline, 39, 40, **40**, 47, **92**

Tangle Falls, **125**
Tangle Ridge, 123
Tatei Formation, 51
Tekarra Lake valley, 113
Terranes of western Canada, 19
  superterranes, 19, 20
Tête Jaune Cache
  stretched conglomerate, 51, **52**
Throne Mountain, 149, **149**
Thrust faults, 39, 40, **40**
  differences from normal faults, 26
  typical form of, 37, **38**
Thrust faults and thrust sheets, 21
  defined, 22
Tintina Trench, 52
Titkana Formation, 147
Tonquin Valley, 11, **98**, 149–52
Trident Range, 152
Trilobites, 11
Tuffa deposits, 77–78, **78**
Tyndall limestone, 3, **4**

Utopia Mountain, 80

Verdant Pass, 149
Victoria Cross Ranges, 137
Vine Creek valley, **54**
Volcanic island arcs, 19

Watchtower Mountain, 11, **91**
Watchtower Valley, 102
Western Canada plate tectonics, **18**
Whirlpool Fault, 149, 150
Whirlpool River, 32
Whistlers (The), 142–44
  view from, **142**, 144
Wisconsinan glaciation, 30, 32, 150
  age of, 32

Yellowhead Lake, 47, **48**
Yellowhead Pass, 32, 46